National Conference on Smart Electronics Engineering the Future of Intelligence (Nc'sefı-2024)

-: EDITORS :-

Dr. Fahimuddin Shaik

Dr. Ch. Nagaraju

Dr. Shaik Karimullah

Dr. C. Venkatesh

TANEESHA
PUBLISHERS

Title : National Conference on Smart Electronics Engineering the Future of Intelligence (Nc'sefi-2024)

Editors : Dr. Fahimuddin Shaik, Dr. Ch. Nagaraju
Dr. Shaik Karimullah, Dr. C. Venkatesh

Edition : First (March, 2025)

ISBN : 9788197680847

Published by

TANEESHA PUBLISHERS | *A Venture by -* **PRACHI DIGITAL PUBLICATION**

Regd. Add.: 254, Khuriyakhatta No. 10, Bindukhatta,
Lalkuan, Nainital - 262402, Uttarakhand, India
Website : www.taneeshapublishers.in
E-mail : taneeshapublishers@gmail.com
Phone : +91 845481 2712, +91 976041 7980

Printed by :
Manipal Technologies Limited, Bengaluru - 560001, Karnataka

PREFACE

It is with great pleasure that we present the Abstract Proceedings of the National Conference on Smart Electronics: Engineering the Future of Intelligence (NC'SEFI-2024), held on 20th April 2024 in hybrid mode, encompassing both online and offline participation. This conference marks a significant gathering of scholars, researchers, industry experts, and students, all united by a shared commitment to exploring the cutting-edge advancements in smart electronics and their pivotal role in shaping the future of intelligence.

The rapid evolution of electronics, fueled by innovations in artificial intelligence, machine learning, and the Internet of Things (IoT), has ushered in a new era of smart systems that are transforming industries and daily life. NC'SEFI-2024 serves as a platform for the dissemination of groundbreaking research, fostering dialogue and collaboration among the brightest minds in the field.

This volume of abstract proceedings encapsulates the diverse range of topics covered at the conference, from novel sensor technologies and embedded systems to advancements in robotics, automation, and energy-efficient electronics. Each contribution reflects the authors' dedication to pushing the boundaries of knowledge and their vision for engineering a smarter, more connected world.

We extend our sincere gratitude to all the participants, speakers, and organizers whose efforts have made NC'SEFI-2024 a resounding success. We also thank the reviewers and editorial team for their diligent work in ensuring the high quality of the submissions included in this compilation. It is our hope that these proceedings will serve as a valuable resource for ongoing research and inspire further innovations in the dynamic field of smart electronics.

As we look toward the future, we are confident that the ideas and collaborations sparked by this conference will continue to drive progress in engineering intelligence, contributing to the betterment of society as a whole.

20th April 2024

Conference Organizing Committee
National Conference on Smart Electronics: Engineering the Future of Intelligence (NC'SEFI-2024)

MESSAGE FROM CHANCELLOR, ANNAMACHARYA UNIVERSITY, RAJAMPET

I am immensely pleased to extend my warmest greetings to all participants of the National Conference on Smart Electronics: Engineering the Future of Intelligence (NC'SEFI-2024). This conference represents a remarkable opportunity for the convergence of knowledge and innovation, bringing together experts from academia and industry to explore the future of smart electronics. At Annamacharya University, we are committed to fostering an environment that encourages pioneering research and the development of solutions that address real-world challenges. I wish all attendees a fruitful and inspiring experience as we collectively work towards engineering a smarter future.

Dr. C. Gangi Reddy
Chancellor,
Annamacharya University,
Rajampet

MESSAGE FROM PRO_CHANCELLOR, ANNAMACHARYA UNIVERSITY, RAJAMPET

It is with great pride that I welcome you to NC'SEFI-2024, a conference dedicated to the exploration of smart electronics and their transformative impact on our world. As we stand on the brink of unprecedented technological advancements, this event serves as a vital platform for the exchange of ideas and collaboration. I am confident that the discussions and presentations held here will pave the way for significant innovations that will shape the future of intelligent systems. My best wishes to all participants and organizers for a successful and productive conference.

Mr. C. Abhishek Reddy
Pro-Chancellor,
Annamacharya University,
Rajampet

MESSAGE FROM THE PRINCIPAL
AITS, RAJAMPET

On behalf of the Annamacharya Institute of Technology and Sciences, Rajampet, I am delighted to welcome all participants to NC'SEFI-2024. This conference is a testament to our commitment to academic excellence and our belief in the power of technology to drive societal progress. The topics covered in this conference are at the forefront of modern engineering, and I am certain that the knowledge shared here will contribute greatly to the advancement of smart electronics. I extend my heartfelt thanks to all contributors and wish everyone an enriching experience.

Dr. SMV. Narayana
Principal, AITS, Rajampet

MESSAGE FROM REGISTRAR, ANNAMACHARYA UNIVERSITY, RAJAMPET

It is an honor to welcome you all to the National Conference on Smart Electronics: Engineering the Future of Intelligence (NC'SEFI-2024). This conference brings together a diverse group of experts and scholars who are passionate about the future of electronics and intelligent systems. I am confident that the insights gained and the collaborations formed here will have a lasting impact on the field. I commend the organizing team for their dedication and wish all participants a successful and enlightening conference.

Dr. Mallikharjuna Rao Nuka
Registrar
Annamacharya University,
Rajampet

MESSAGE FROM HEAD OF THE DEPARTMENT

I am thrilled to see NC'SEFI-2024 come to fruition, offering a platform for innovative minds to discuss and explore the latest advancements in smart electronics. As the Head of the Department of Electronics and Communication Engineering, I am proud of the contributions our department has made towards this event. The research and ideas presented at this conference will undoubtedly inspire further developments in the field, driving us towards a future where intelligent systems enhance every aspect of life. I extend my best wishes to all participants and look forward to the exciting discoveries that will emerge from this conference.

Dr . CH. Nagaraju
Professor & HOD -ECE

Novel Vehicle Security System Using Driver's License and Fingerprint Automation

R L B R Prasad Reddy
Associate Professor, Department Of ECE
Srinivasa Institute Of Technology And Science, Kadapa

U. Hema Narasimha Vardhan
Student, Department Of ECE
Srinivasa Institute Of Technology And Science, Kadapa

Abstract

This paper introduces a pioneering vehicle security system that combines the authentication of driver's licenses and fingerprint biometrics to significantly enhance the protection against vehicle theft. The system is built around an Arduino controller, which interfaces seamlessly with an RFID (RadioFrequency Identification) reader and a fingerprint sensor. The RFID reader constitutes the first level of security, requiring the driver's license to be equipped with an embedded RFID tag for access to the vehicle. Furthermore, the second level of security is bolstered by the fingerprint sensor, which scans the driver's fingerprint for authentication. Only when both levels of security are successfully verified does the system grant access to the vehicle. This novel approach to vehicle security offers robust protection by ensuring that only authorized individuals with both the driver's license and the registered fingerprint can start and operate the vehicle, effectively reducing the risk of theft and unauthorized use.

Detection of Coronary Arterial Disease Using E-Nose Technology

Nithya Kalyani K
(Biomedical Department) Vel Tech Multi Tech Dr Rangaraja
Dr Sakunthala Engineering College
Raj Kumar S
(Biomedical Department) Vel Tech Multi Tech Dr Rangarajan
Dr Sakunthala Engineering College
Surya B
(Biomedical Department) Vel Tech Multi Tech Dr Rangarajan
Dr Sakunthala Engineering College
Sulaiman A
(Biomedical Department) Vel Tech Multi Tech Dr Rangarajan
Dr Sakunthala Engineering College

Abstract

This Approach dealing with the application of electronic nose technology for the detection of Coronary Arterial Disease by analysing Volatile Organic Compounds present in a patient's breath. Cardiovascular diseases, including coronary arterial disease, remain a leading cause of morbidity and mortality globally. To detect this disease, semiconductor sensors like MQ-4, MQ-7 are designed to sense the specific biomarkers like acetone, benzene, toluene which is exhale from the patient. The importance of this research is the way it can transform cardiovascular diagnostics. This approach also deals with signal processing and classification of CAD samples from normal person breath samples by using Machine learning Algorithm.

This paper explores the application of deep learning techniques for predicting the phase of brain tumors, aiming to assist practitioners in assessing patients' conditions and forecasting potential anomalies. Leveraging deep learning facilitates correlating patient data with similar cases, offering a valuable tool for remote healthcare, particularly pertinent during the ongoing pandemic The study focuses on employing this deep learning approach to classify brain images depicting various tumor types: meningioma, glioma, and pituitary. Evaluation utilizing standard metrics within the Matlab environment reveals confusion matrices indicating accuracy of 96. 8%. Comparative analysis with Convolutional Neural Network (CNN) and Artificial Neural Network (ANN) models further underscores the effectiveness of the proposed deep learning method.

Voice Assistant using Python for Desktop

Harinni. R
Veltech Multitech Engineering College, Avadi, Chennai
Nijeth Kumar
Veltech Multitech Engineering College, Avadi, Chennai
Gokulnath. J
Veltech Multitech Engineering College, Avadi, Chennai
Brindha. B
Veltech Multitech Engineering College, Avadi, Chennai

ABSTRACT

This abstract introduces the development of a desktop voice assistant leveraging Python and Natural Language Processing (NLP) techniques. The assistant will mimic popular voice assistants by enabling hands-free interaction with computers through spoken commands.

The system will leverage various Python libraries and frameworks, including speech recognition, natural language processing, and automation tools, to achieve its functionality. It utilizes Python libraries like Speech Recognition, NLTK/spacy, and pyttsx3. Additionally, API integration and customization features can be incorporated for expanded functionalities and user-specific preference.

Key functionalities include Speech recognition for converting spoken commands to text. NLP techniques to comprehend user intent and extract entities from commands. Text-to-speech for vocal responses from the assistant. Action execution to fulfil user requests, such as launching applications, conducting web searches, or controlling media playback.

In conclusion this desktop voice assistant, built with Python and

NLP, presents a user-friendly and efficient way to interact with computers through voice commands, promoting improved user experience, accessibility, and productivity.

High Speed Gate Level Syynchronous Full Adder Designs

Syed Jaffar Ali
Dept of ECE, Annamacharya institute of technology and sciences: Kadapa
Vuntla Yugandhar Reddy
Dept of ECE, Annamacharya institute of technology and sciences: Kadapa
Pulla Bhanu Prakash Reddy
Dept of ECE, Annamacharya institute of technology and sciences: Kadapa
Vasa Shanmuka Srinivas
Dept of ECE, Annamacharya institute of technology and sciences: Kadapa
Y. Raghupathi Naidu
Dept of ECE, Annamacharya institute of technology and sciences: Kadapa

Abstract

Addition forms the basis of digital computer systems. Three novel gate level full adder designs, based on the elements of a standard cell library are presented in this work: one design involving **XN**OR and **m**ultiplexer gates (XNM), another design utilizing **XN**OR, **A**ND, **I**nverter, **m**ultiplexer and **c**omplex gates (XNAIMC) and the third design incorporating **X**OR, **A**ND and **c**omplex gates (XAC). Comparisons have been performed with many other existing gate level full adder realizations.

Based on extensive simulations with a 32-bit carry-ripple adder implementation; targeting three process, voltage and temperature (PVT) corners of the high speed (low V_t) 65nm STMicroelectronics CMOS process, it was found that the XAC based full adder is found to be delay efficient compared to all its gate level counterparts, even in comparison with the full adder cell available in the library. The XNM based full adder is found to be area efficient, while the XNAIMC based full adder offers a slight compromise with respect to speed and area over the other two proposed adders.

Sentiment Driven Equity Market Forecasting

Haritha K
Veltech Multitech Engineering College, Avadi, Chennai
Vignesh S
Veltech Multitech Engineering College, Avadi, Chennai
Mukesh G
Veltech Multitech Engineering College, Avadi, Chennai
Jayanth T K
Veltech Multitech, Engineering College, Avadi, Chennai

ABSTRACT

This paper presents an overview of the literature on the integration of sentiment research into stock price forecasting algorithms. It looks at social media, news articles, and financial reports as sources of sentiment data and covers a variety of sentiment analysis methodologies, such as machine learning and lexicon-based approaches. The effectiveness of sentiment analysis across a range of assets and market circumstances is reviewed, and techniques for incorporating it into forecasting models like sentiment-based feature engineering and event detection are also covered. This paper provides useful information about applying sentiment analysis to increase the accuracy of stock price forecasts, as well as ideas for future study in this field.

Brain Tumor Detection Using Deep Cnnfusion With Istm

P. V. Kusuma
Department of ECE, JNTUA College of Engineering Pulivendula.
B. Harini
Department of ECE, JNTUA College of Engineering Pulivendula.
I. Surya Prakash
Department of ECE, JNTUA College of Engineering Pulivendula.
K. Tejaswini
Department of ECE, JNTUA College of Engineering Pulivendula.
S. Joshna
Department of ECE, JNTUA College of Engineering Pulivendula.

ABSTRACT

Brain tumors are recognized as severe illnesses wherein the precision of images plays a crucial role. Accurately identifying tumors is essential for precisely pinpointing the affected area and thereby reducing the mortality rate. Consequently, understanding the hidden patterns becomes imperative for an improved diagnosis and image quality. However, achieving accurate diagnoses across various lesion cases poses a significant challenge.

To address the limitations of existing methods, we introduce the Gaussian Weighted Deep Convolutional Neural Network with LSTM (GW Deep CNN-LSTM) for the automatic detection of brain tumor patients from their tumor images. The GW Deep CNN-LSTM technique comprises multiple layers. Initially, brain MR images are retrieved from the designated database, and image preprocessing is carried out using a Gaussian weighted non-local mean filter to eliminate noisy pixels. Subsequently, segmentation is implemented using Hartigan's segmentation method to partition the image into similar regions.

Following segmentation, feature extraction is conducted to extract more informative features, including texture, color, and intensity, from the segmented image. Finally, the classification of brain MR images is accomplished through Long Short-Term Memory (LSTM). This classification process enables the identification of the input image as normal or tumor with higher accuracy. Notably, GW Deep CNN-LSTM demonstrates superior performance in disease detection accuracy while minimizing the time and error rate.

Advanced Opensource Raspberry Pi-Powered Digital Photo Frame Using Iot

P. Vijaya
Assistant professor, Depaartment of ECE,
AITS Kadapa.
G. Mahidhar Reddy
Student, Department of ECCE,
AITS Kadapa.

ABSTRACT

As smartphones evolving day by day, Images have become so important in our lives, and more and more are taken every day. Due to our busy schedule we can't able to see all the photos often. If it's in our smartphone, when we find sometime we can see them but as years passed we have pile of pictures and all of them hardly we remember.

Also we have photo frames in our home which were something to cherish when we saw them, sometimes we may feel that we can have some more best photos on the wall. More photo frames in the home/office leads to difficulty of maintenance and prone to damage. Hence we came up with a solution called Advanced open source digital frame. Digital picture frames are really nice because they let you enjoy your photos without having to print them out. Plus, adding and removing digital files is a lot easier than opening a traditional frame and swapping the picture inside when you want to display a new photo.

Even so, it's still a bit of overhead to remove your SD card, USB stick, or other storage from a digital picture frame, plug it into your computer, and copy new pictures onto it. An easer option is a digital picture frame that gets its pictures over WiFi, for example from a cloud service. We are going to add motion sensor and light ambient senor for effective control of display.

Keywords: IOT, Digital Photo Frame, Raspberry pi, Cloud services, Opensource

Analysis of the Effect of Slots and Dgs on Rectangle Patch Antenna

S. Kavya
Department of ECE, JNTU College of Engineering,
Pulivendula, AP, India, singareddykavya123@gmail.com
K. Divya
Department of ECE, JNTU College of Enginnering,
Pulivendula, AP, India, kalimisettydivya@gmail.com
B. Gowtham
Department of ECE, JNTU College of Engineering,
Pulivendula, AP, India, bgowtham2002@gmail.com
V. Prasanna Kumar
Department of ECE, JNTU College of Engineering,
Pulivendula, AP, India, vaddiprasannakumar359@gmail.com
P. Mohith Kumar Achari
Department of ECE, JNTU College of Engineering
Pulivendula, AP, India, pirangimohithkumar@gmail.com
R. Ramana Reddy
Professor, Department of ECE, JNTU College of Engineering
Pulivendula, AP, India, profrrreddy@yahoo.co.in

ABSTRACT

A rectangle patch antenna incorporated with a defective ground structure is proposed. The proposed antenna's overall dimensions are 36x39x1. 6 mm^3. The dimensions of patch are 25x32 mm^2. The rectangle patch antenna with pentagonal and hexagonal slots was designed. Both the hexagonal slotted patch antenna and pentagonal slotted patch antenna yielded a bandwidth of 120MHz with a peak gain of 7. 6dBi.

Further to improve the bandwidth and gain, defective ground structure is incorporated. Incorporation of DGS enhanced the bandwidth to 650MHz (5. 47-6. 11GHz) with a peak gain of 15. 4 dBi. The proposed antenna is suitable for C-band applications. Simulations were carried out using Ansys HFSS software.

Keywords - Defective ground structure, C-band, rectangle patch, HFSS.

Automatic Industrial Fault Detection and Iot Based Remote Monitoring

D. Siva Dharani
Department of ECE, SRTS, Kadapa,
sivadharanidasareddygari@gmail.com
C. Swathi
Department of ECE, SRTS, Kadapa,
chirapaswathi@gmail.com

ABSTRACT

Internet of Things (IoT) technology is advancing quickly. Many home and commercial IoT applications are automated. Currently, monitoring and automation refer to the ability for a user to control something from every where in the world, rather than just when they are physically present in a given location. IoT, or things that are embedded with sensors, software, electronics, and network connectivity, refers to a network of actual objects that can gather and share data. Controlling physical goods remotely through the Internet is another use for IoT.

Different sensors are used to monitor industrial features including temperature, variations in current and voltage parameters, gas, etc. By considering the potential threats to the normal operation of the industry equipment, these variables were carefully selected. The sensors used here are Flame sensor LM393, Current Sensor ACS 712ELCTR- 05B-T, Voltage sensor, Ultrasonic Sensors HC-SR04, Temperature sensor DHT11, Gas sensor MQ9, These Sensors will collect their respective data and send the data to Node MCU ESP which also acts as a wi-fi module Using the IoT idea, the paper is being controlled and monitored via fault detection, status monitoring, and alarm/alert generation when a fault arises and promptly reports it to the appropriate service Centre. This attempts to improve productivity, regulate access, and produce high-quality industrial goods.

Keywords— IoT, Node MCU, Sensors, WIFI Module.

Combined Use of a Bilateral and Median Filter to Suppress Gaussian Noise in Images

B. Bhavana
Department of ECE, SRTS, Kadapa, bhavanabadvel@gmail.com
K. Dhanaganga
Department of ECE, SRTS, Kadapa, kunigalladhanaganga@gmail.com

ABSTRACT

The paper is devoted to the problem of processing images with Gaussian noise. The method proposes a new way of denoising. This algorithm can effectively clean the image from noise and keep the edges of objects precise. The research was conducted in MATLAB R2019b, the quality of the proposed filter was tested using a peak signal-to-noise ratio and a structural similarity index. This method can de-noise images more efficiently in comparison with most existing methods. The proposed method de-noise better by 1-6 dB than comparable methods in the work. It can be used in applications, image processing, or as a precleaning program for an image recognition neural network.

Keywords- bilateral filtering, median filtering, peak signal-to-noise ratio, Gaussian noise, image denoising filter

A Cryptographic System for Bidding Privacy in Blockchain Auctions

Haritha
Department of AI&DS, Vel Tech Multi Tech
K. Gowseik V
Department of AI&DS, Vel Tech Multi Tech
Dr. Rangarajan
Dr. Sakunthala Engineering College, Dr. Rangarajan Dr. Sakunthala Engineering College
Avadi
Department of AI&DS, Vel Tech Multi Tech Dr. Rangarajan Dr. Sakunthala Engineering College, Avadi
Sanjay C
Department of AI&DS, Vel Tech Multi Tech Dr. Rangarajan Dr. Sakunthala Engineering College, Avadi
Shafeek Ansari A
Department of AI&DS, Vel Tech Multi Tech Dr. Rangarajan Dr. Sakunthala Engineering College, Avadi

Abstract

Blockchain-based auction platforms have become more and more popular in business, providing answers to security and trust issues that come with conventional bidding sites. This paper investigates the novel cryptographic scheme used in "Bid Confidentiality in Blockchain Auctions. " The framework leverages the hash-chained and decentralized features of blockchain technology to tackle crucial problems related to the integrity of bidding processes.

Traditional systems frequently struggle with third-party channel trust, a problem that blockchain's built-in security eliminates. By ensuring bid confidentiality, the cryptographic design safeguards user identities and bid data. Strong cryptographic protocols notably support the entire auction lifecycle, from management to bidding procedures. The framework creates a safe, transparent, and effective environment for blockchain auctions by integrating smart contracts, secure data storage, and cryptographic verification techniques. Using this perspective on cryptography, the paper adds to improving the confidentiality, honesty, and general reliability of blockchain-based auction systems.

A Cyber Security in Dc Microgrids for Detecting False Data Injection by Fuzzy Controller

Muqthiar Ali Shaik
Asst. Professor, Dept. of EEE, AITS, Rajampet.
G. Naveen Kumar
UG Students, Dept. of EEE, AITS, Rajampet
L. Teja Sree
UG Students, Dept. of EEE, AITS, Rajampet.
S. Maqsood Ahamad
UG Students, Dept. of EEE, AITS, Rajampet.
M. Hemanth Naga Sai
UG Students, Dept. of EEE, AITS, Rajampet.

Abstract

With the growing integration of digital technologies in power systems, the vulnerability to cyber-attacks has become a critical concern. False data injection attacks, in particular, pose a significant threat to the reliable operation of DC micro grid clusters. This research addresses the detection of false data injection cyber-attacks in the control system of multiple DC micro grid clusters.

The existing system employs a Proportional-Integral-Derivative (PID) controller in the control block to regulate the micro grid clusters. However, traditional controllers like PID are susceptible to malicious manipulations through false data injection attacks, compromising the stability and performance of the micro grid clusters. In response to this vulnerability, a proposed system is introduced, incorporating a Fuzzy controller in the control block. Fuzzy logic controllers offer robustness and adaptability, making them more resilient to cyber threats.

The integration of the Fuzzy controller aims to enhance the cyber-physical security of the DC micro grid clusters by improving the system's ability to detect and mitigate false data injection attacks. The study evaluates the performance of both the existing PID-based system and the proposed Fuzzy controller- based system under normal operating conditions and in the presence of simulated false data injection attacks. Comparative analyses are conducted to highlight the advantages of the Fuzzy controller in terms of detecting and responding to cyber- attacks while maintaining the desired stability and performance of the DC micro grid clusters.

Keywords— DC Microgrid, PID Controller, fuzzy controller, cyber security etc.

High Protection Locker Security System Using Live Image and Voice Authentication

V NEELIMA

Department of ECE Srinivasa Institute Of Technology And Science, Kadapa neelimav@nda.edu.in

T. DIVYA

Department of ECE Srinivasa Institute Of Technology And Science, Kadapa divyalokeshds@gmail.com

Abstract

we need security to keep us and our valuable things safely. As the technology is developing rapidly we should also upgrade the security protection to prevent ourself from illegal handling and losing of our things. This high protection locker security system helps us to keep our valuable things in banks, homes and hotels securely. The security system we are using now can be easily accessible by unauthorized person and theft because of the advanced technology. To prevent this we are using high protection locker system which uses image recognition and voice authentication Keywords: Arduino uno, Locker, Node mcu, security, python, image and voice I INTRODUCTION Here we are using web camera and Node MCU for identifying the person in python and his voice (i. e., Authentication Key). We can compare with the registered images of person if it is matches with the register image and based on the voice command the locker will open otherwise it will not open and stores the image of the intruders.

Keywords— Arduino uno, Locker, Node mcu, security, python, image and voice

Ecg Signal Extraction Method Based on Singular Value Selection and Wavelet

K. Naga Sankar Reddy
S V College of Engineering, Balaji Nagar, Kadapa, A. P. –516003,
knsreddy.1987@gmail.com, Mobile no : 9550668898
Ponala Prasad
S V College of Engineering, Balaji Nagar, Kadapa, A. P. –516003
ponalaprasad86@gmail.com, Mobile no : 8688264737

ABSTRACT

Aiming at the noise of the received ECG signal, a extraction method of ECG signal based on singular value selection and wavelet is proposed. The singular value decomposition on the ECG signal is performed firstly, and the ECG signal component corresponding to each singular value is obtained. Then the signal component corresponding to the maximum singular value is used to calculate cross-correlation coefficients with other components. The cumulative contribution rate of singular values is combined to determine the number of singular values for ECG signal reconstruction. The wavelet threshold de-noising method is used to de-noise the final determined signal components. Finally, the de-noising ECG signal is obtained by reconstructing the signal components. The experimental results show that the method can suppress noise and extract signal effectively, and it has a good noise reduction effect compared with wavelet threshold method.

Keywords- ECG signal, signal extraction, singular value decomposition, wavelet

Empowering Women's Safety with Smart Iot Technology: a Robust Protection System

A. Guru Chandana
Department of ECE, SRTS, Kadapa
K. Ankireddy
Department of ECE, SRTS, Kadapa

Abstract—Nowadays, women's safety has emerged as one of the world's most pressing issues, and crime against women has significantly increased nationwide. The frequency of auto theft and accidents has increased significantly as compared to the previous decade. The value of a human life exceeds all other values, and it is more crucial to seek assistance when you do so than to provide a helping hand. Every day, women from all walks of life struggle to maintain their safety and defend themselves. In the presence of insensitive guys who regularly harass, assault, and violate the dignity of women, and in front of their wandering eyes. Public transport, especially public places, has become hunter territory. For these atrocities that women undergo current scenario A clever wearable safety tool for girls primarily based totally at the Internet of Things is being implemented. It consists of a Arduino, Camera and a buzzer, and a button to activate the service. This unit is very remote and with the click of a button, it is activated by the victim when attacked, retrieves their current location and captures images of the victim Attacker via Esp32 Cam Module. A predetermined contact number will receive both the taken image and the location link. It monitors through the victim's smartphone and avoids the usage of additional hardware or modules, keeping the device compact. Live video streaming is via webcam and live Position is obtained by GPS module. This paper makes a collision avoidance and detection systems. There are numerous initiatives being taken with the goal of saving lives. The proposed study intends to develop an IoT-based security system that prioritises protecting women.

Glucose Content Analysisusing Image Processing and Machine Learning Techniques

K. Praveen kumar reddy
Department of ECE SRTS, Kadapa
A. Nageswareddy
Department of ECE SRTS, Kadapa

Abstract: Technology is constantly evolving to make it easier for people to work in biomedical research and technology daily. The glucose level checking system can use a urine analyzer detector as a color reader of the urine strip. This work aims to analyse glucose levels based on digital picture identification using the MATLAB application for patient glucose data processing. Injections are joint for diabetic people to control their blood sugar levels. Repeated injections might cause minor physical harm to the body that can hamper the immune system's ability to fight against pathogens. Numerous research has concentrated on non-invasive glucose-based testing, namely using urine. This study was created using image processing to examine the non-invasive glucose testing procedure. The noise is cleaned up using a Gaussian filter and histogram-based feature extraction for picture database extraction. Support vector machines classify data using a 70% training and 30% testing process. The SVM classification results had an accuracy of 85% and time processing of 0. 5 seconds. In making medical decisions, it is possible to consider the effects of diabetes, pre-diabetes, and non-diabetes.

Digital Agrarian Renaissance: Revolutionizing Farming Practices and Livelihoods in Agroworld

M. Swetha
Artificial intelligence and data science, Anna university
vel tech multi tech engineering college, Anna university
chennai, India, swetham@veltechmultitech. org
M. Srinivasan
Artificial intelligence and data science, Anna university
vel tech multi tech engineering college, Anna university,
srinichennu11@gmail.com
V. Viknesh
Artificial intelligence and data science, Anna university
vel tech multi tech engineering college, Anna university
chennai, India, jeyaviknesh@gmail.com
K. Farhan shahul
Artificial intelligence and data science, Anna university
vel tech multi tech engineering college, Anna university
chennai, India, farhansha1710@gmail.com

Abstract: The integration of mobile internet stands poised to revolutionize agricultural commerce, empowering farmers to directly engage with consumers and food processing industries, thereby facilitating the sale of their produce. This paper proposes a user-friendly mobile application aimed at delivering market information to farmers efficiently. With its intuitive interface and support for native languages, the application seeks to streamline transactions for farmers, treating them as both sellers and buyers. The central aim is to empower farmers by providing them with a platform to effectively market and trade their agricultural goods and products. Leveraging market data from sources like data. gov. in enables the system to maintain oversight of selling and buying prices. In order to simplify the browsing experience, the application incorporates multiple filters to accommodate the diverse range of available products. By addressing the common challenges faced by farmers in selling their produce, this system offers a convenient and streamlined platform. It ensures equitable and transparent transactions, allowing farmers to sell their goods at fair prices. Likewise, consumers benefit from a wide array of products and have the option to filter based on their preferences, including price and location. The primary objective of the system is to cater to the needs of both producers and consumers, facilitating fair and transparent transactions within the agricultural sector.

Keywords—component, Mobile Application, Food Processing Industry, Agriculture

Iot-Based School Children Transportation Safety System

G. NANDINI
Department OF ECE, SRTS, Kadapa, gubalanandini123@gmail.com
P. TARAKA RAJASEKHAR
Department OF ECE, SRTS, Kadapa, rajasekharp638@gmail.com

Abstract

The "paper offers a comprehensive solution to enhance the safety and security of students during their school bus commute. A Python- installed PC equipped with a camera and OpenCV is employed for real-time face recognition of students as they board and disembark from the school bus. The system incorporates alcohol detection capabilities for the driver. The bus's ignition system will remain locked, preventing the bus from starting . Furthermore, the system enhances student safety by implementing accident detection and fire detection mechanisms. Simultaneously, an email notification is dispatched to the school administration, allowing immediate action. This paper represents a significant advancement in ensuring the safety and security of students during their school bus commute, promoting proactive monitoring and immediate responses to critical situation

IoT Enabling Night Patrolling Robot for Women Safety

Policherla Lavanya,
Assistant Professor, Department of ECE,
Sri Venkateswara College of Engineering, Kadapa
Vadla Lakshmi Swetha
UG Student, Department of ECE,
Sri Venkateswara College of Engineering, Kadapa

ABSTRACT

The paper "IoT Enabling Night Patrolling Robot for Women Safety" aims to address the pressing need for enhanced safety measures for women in urban areas. In today's world, ensuring women's security is paramount, and leveraging advanced technology is crucial. This paper focuses on developing a simple and cost-effective safety gadget that can assist women in times of danger. The gadget includes a switch, Nodemcu Board, Wi-Fi module, buzzer, motor driver, high-definition night vision camera, and pushbuttons. When activated, the device tracks the woman's location using GPS and sends emergency messages to pre-registered mobile numbers and the police control room via the Wi-Fi module. Additionally, it sends the current GPS location to an ambulance every 10 seconds as email alerts. The compact and portable design of the device makes it easy to carry, and the use of sophisticated components ensures its accuracy and reliability. Overall, this paper aims to leverage IoT technology to create a simple yet effective safety solution for women. By combining GPS tracking, Wi-Fi communication, and emergency alert mechanisms, the device has the potential to significantly improve women's safety in urban environments.

Monitoring of Soil Nutrients, Humidity Levels and Water Levels for Different Lands Using Iot

B. Saroja
Professor& HOD, ECE Siddartha Institute of Science and Technology, India, boda.saroja@gmail.com
Kalathuru Sethavi
Department of ECE Siddartha Institute of Science and Technology, India, sethavikalathuru@gmail.com
Bandaru Reddy Chandu
Department of ECE Siddartha Institute of Science and Technology, India, reddychandubandaru@gmail.com
Chamarthi Rajesh
Department of ECE Siddartha Institute of Science and Technology, India, Chamarthirajesh2@gmail.com
Putta Reddy Prasanna
Department of ECE Siddartha Institute of Science and Technology, India, Puttareddyprasanna1234@ gmail.com
Rajugari Sai Dheeraj Reddy
Department of ECE Siddartha Institute of Science and Technology, India, rajugarisaidheerajreddy@gmailgmail.com

Abstract

The "Monitoring of Soil Nutrients, Humidity Levels and Water Levels for Different Lands Using IoT" introduces a ground-breaking advancement in agricultural technology by combining the Internet of Things (IoT) with soil nutrient and water level analysis. This system takes advantage of interfacing Arduino with an NPK sensor and a soil moisture sensor to carefully measure NPK values and moisture content in the soil. The data collected from these sensors is then prominently displayed on an LCD screen for immediate reference. The GSM will able to upload the sensor readings to a webserver for remote monitoring and analysis. In this innovative approach, the NPK sensor and soil moisture sensor work together to constantly monitor the soil's nutrient composition and moisture levels. This data gets transmitted to the Arduino microcontroller, which processes and interprets the information. By utilizing present threshold values, the Arduino controls three different water pumps for nitrogen (N), phosphorus (P), and potassium (K) irrigation, based on the specific nutrient deficiencies detected. Furthermore, a standard pumping motor is used to manage soil moisture levels according to predefined thresholds. This responsive and dynamic system ensures that crops receive tailored and optimal irrigation, promoting healthy growth and enhanced yield potential.

Keywords: Smart soil analysis; Laboratory soil analysis; Land use patterns; soil depth; soil properties, moisture levels, nutrient values., Arduino with NPK sensor.

Multiple Disease Prediction System Using Machine Learning

K. Kondamma
Department of Electronics and Communication Engineering,
JNTUA College of Engineering Pulivendula.
N. Sireesha
Department of Electronics and Communication Engineering,
JNTUA College of Engineering Pulivendula.
D. Niharika
Department of Electronics and Communication Engineering,
JNTUA College of Engineering Pulivendula.
D. Purushotham
Department of Electronics and Communication Engineering,
JNTUA College of Engineering Pulivendula.
M. Mani Kumar Reddy
Department of Electronics and Communication Engineering,
JNTUA College of Engineering Pulivendula.

Abstract

Machine learning (ML) refers to the science and engineering of artificially intelligent systems, providing them with the capability to learn without being explicitly programmed. In recent years, ML in the healthcare domain has made great advancements in the early predictions of many critical illnesses.

While there have been significant contributions to single disease prediction systems (like one for heart disease prediction, one for diabetes, and so on), there lacks the existence of an integrated regulatory framework that performs the prediction of multiple diseases under a single interface. This paper proposes a novel "Multiple-disease Prediction System" for the early-stage prediction of Heart Diseases, Parkinson's Disease, Breast Cancer, and Diabetes using efficient Machine Learning algorithms.

The novelty of this work lies in the implementation of multiple ML classifiers for each prediction system, followed by their performance evaluation and comparison based on their respective test accuracies. From experimental results, it was observed that Logistic Regression (LR) and Support Vector Machine (SVM) outperformed Decision Tree and Random Forest classifiers in all cases. The highest reported accuracy for LR was 77% for the diabetes model, while SVM obtained accuracy values of 80%, 92%, and 97% for heart disease, Parkinson's disease, and breast cancer models, respectively.

Railway Track Fault Detection Robot

K. Sandeep Kumar
Associate Professor, Dept of ECE, Siddartha Institute of Science and Technology, Puttur, India.
M. Keerthi
Dept of ECE, Siddartha Institute of Science and Technology, Puttur, AP, India.
P. Munikishore
Dept of ECE, Siddartha Institute of Science and Technology, Puttur, AP, India.
D. Mohith
Dept of ECE, Siddartha Institute of Science and Technology, Puttur, AP, India.
N. Manogna
Dept of ECE, Siddartha Institute of Science and Technology, Puttur, AP, India.
N. Nandeeswar Reddy
Dept of ECE, Siddartha Institute of Science and Technology, Puttur, AP, India.

Abstract

Railway transportation requires constant inspections and immediate maintenance to ensure public safety. Traditional manual inspections are not only time consuming, and expensive, but the accuracy of defect detection is also subjected to human expertise and efficiency at the time of inspection.

Computing and Robotics offer automated IOT based solutions where robots could be deployed on rail-tracks and hard to reach areas, and controlled from control rooms to provide faster inspection. In this paper, robot will go on railway track and ultrasonic sensor is used for detecting the defect in track. If any defect is detected it will send the location to mobile using GSM. It will track the location using GPS.

The system provides ultrasonic sensor for defect detection, storage of information that consist of location of the defected railway tracks, and robot localization within a range of 3-6 inches. The proposed system utilizes state of the art Machine Learning system and obtained from the tracks to classify them as normal or suspicious. Such locations are then marked and more careful inspection can be performed by a dedicated operator with very few locations to inspect.

Suthenticated Access Control for Vehicleignition System by Driver's License and Fingerprint Technology

K. Sandeep Kumar
Associate Professor, Dept of ECE, Siddartha Institute of Science and Technology, Puttur, AP, India, krishnam.sandeep@gmail.com
C. Sravanthi
Dept of ECE, Siddartha Institute of Science and Technology, Puttur, AP, India, chengannagarisravanthi595@gmail.com
N. Ramya
Dept of ECE, Siddartha Institute of Science and Technology, Puttur, AP, India, ramyanagineni0@gmail.com
K. Rohith Kumar
Dept of ECE, Siddartha Institute of Science and Technology, Puttur, AP, India, katarirohithkumar2004@gmail.com
T. Sai Vardhan
Dept of ECE, Siddartha Institute of Science and Technology, Puttur, AP, India, saivardhan3668@gmail.com
V. Srinuvasullu
Dept of ECE, Siddartha Institute of Science and Technology, Puttur, AP, India, vutisrinivasulu@gmail.com.

Abstract

Fingerprint identification is one of the most popular and reliable personal biometric identification methods. By using this biometric authentication, we can prevent non-license from driving. While issuing the license, the specific person's fingerprint is to be stored in the RFID card. It tells to place his/her fingerprint, and if the fingerprint is valid it sends message to the person to access the vehicle. If fingerprint is not valid it also sends the message that fingerprint is denied. The same automobile should have the facility of fingerprint reader device. After passing all authentications, the vehicle will be ignited. The vehicle will not be ignited, if any one of the authentications fails and will not proceed to the next step. GSM (Global System for Mobile Communications), it is used to get the alerts messages through mobile. GPS (Global Positioning System), is a satellite-based navigation system that allows users to determine their precise location and track the vehicular movement.

Encryption for DNS Attacks

Hemanand Chittapragada
Asst. Prof., Computer Science & Engineering, Madanapalle Institute of Technology & Science, Madanapalle, India.
Yadiki Akhil Kumar Reddy
UG Scholar, Computer Science & Engineering, Madanapalle Institute of Technology & Science, Madanapalle, India
Paradesi Om Prakash
UG Scholar, Computer Science & Engineering, Madanapalle Institute of Technology & Science, Madanapalle, India
Poojari Bhanu Prakash
UG Scholar, Computer Science & Engineering, Madanapalle Institute of Technology & Science, Madanapalle, India.
Yeggidi Suresh Varma
UG Scholar, Computer Science & Engineering, Madanapalle Institute of Technology & Science, Madanapalle, India

Abstract

The primary objective of this paper is to use a DNS encryption to provide a safe communication between the client and a DNS server. Initializing the client, producing key pairs, and setting up a secure socket to communicate are important processes. The procedure entails the creation of requests for DNS information and encryption answers after a secure public key exchange among the client and DNS server. The client parses the data it has received, decrypts it, and shows the user the results. The primary objective of this paper is to utilize DNS encryption to ensure secure communication between the client and a DNS server. The process begins with initializing the client, which involves configuring the environment to support secure DNS transactions and installing necessary cryptographic libraries.

Key pair generation follows, where the client creates a public key and a private key essential for encrypting and decrypting DNS messages. Establishing a secure communication channel is the next crucial step, involving the setup of a secure socket, often using protocols such as TLS (Transport Layer Security) to ensure that all data transmitted between the client and the DNS server is encrypted and protected from interception. Once the secure channel is established, the client can create requests for DNS information.

Keywords: DNS, Spoofing, Poisoning, Encryption, Dos

Pneumonia Detection Using Cnn Algorithm

V. Nagoorbee
Student, Dept. of ECE, JNTUACEP, AP, India,
venuthurlanagoorbee049@gmail.com
G. Sai Chaithanya
Student, Dept. of ECE, JNTUACEP, AP, India,
saichaitanya1214@gmail.com
P. Usha Sree
Student, Dept. of ECE, JNTUACEP, AP, India,
poolaushasree.royal@gmail.com
Y. Abhishek Reddy
Student, Dept. of ECE, JNTUCEP, AP, India,
abishekreddyas286@gmail.com
S. Varalakshmi
Assistant Professor, Dept. of ECE, JNTUACEP, AP, India,
varalakshmi1190@gmail.com

Abstract

Pneumonia is form of a respiratory infection that affects the lungs. In these acute respiratory diseases, human lungs which are made up of small sacs called alveoli which is air in normal and healthy people, but in pneumonia these alveoli get filled with fluid or pus.

Chest X-rays are used for recognizing the pneumonia infection. Accordingly, to perceive the pneumonia radiologists needs to see the Chest X-rays and he/she wants to revive the expert precisely. The essential objective of this paper is to recognize if the individual has pneumonia or not with high accuracy using Convolutional Neural Network.

Performance of a CNN model over the testing data set is calculated after the completion of the training of the model and the performance is calculated in the following six metrics: accuracy, precision, sensitivity or recall, F1 score, the area under the curve (AUC), and specificity.

Multi-Classification Cnn Model for Brain Tumor Mr Images By Using Resnet-50

Murali Krishna Atmakuri
Department of Electronics and Instrumentation Engineering, Faculty of Engineering and Technology, Annamalai University
Annamalai Nagar, India, muralikrishna.atmakuri80@gmail.com
Ganesh Ram A
Department of Instrumentation Engineeringline, Madras Institute of Technology, Anna University Chennai, India, agram72@gmail.com
V. V. K. D. V. Prasad
Department of Electronics and Communication Engineering
Gudlavalleru Engineering College Gudlavalleru, India,
varrevkdvp@rediffmail.com

Abstract

This paper explores the application of deep learning techniques for predicting the phase of brain tumors, aiming to assist practitioners in assessing patients' conditions and forecasting potential anomalies. Leveraging deep learning facilitates correlating patient data with similar cases, offering a valuable tool for remote healthcare, particularly pertinent during the ongoing pandemic. Digital image digitization and intelligent analysis expedite initial impressions, streamlining manual computations and cross verifications. Deep learning's efficacy in supervised classification of intricate patterns is noteworthy, allowing for the development of fully automated methods using existing toolboxes in simulation environments like Matlab. The study focuses on employing this machine learning approach to classify brain images depicting various tumor types: meningioma, glioma, and pituitary. Evaluation utilizing standard metrics within the Matlab environment reveals confusion matrices indicating accuracy ranging from 89% to 95%. Comparative analysis with Convolutional Neural Network (CNN) and Artificial Neural Network (ANN) models further underscores the effectiveness of the proposed deep learning method.

Brain Tumor Detection Using the CnnAlgorithm With Adam Optimizer

Dr. Shaik Taj Mahaboob
Department of Electronics and Communication Engineering,
JNTUA College of Engineering Pulivendula
B. Gnapika
Department of Electronics and Communication Engineering,
JNTUA College of Engineering Pulivendula
P. Pujitha
Department of Electronics and Communication Engineering,
JNTUA College of Engineering Pulivendula
P. Saraswathi
Department of Electronics and Communication Engineering,
JNTUA College of Engineering Pulivendula
D. Trivikram Chakradhar
Department of Electronics and Communication Engineering,
JNTUA College of Engineering Pulivendula

Abstract

Nowadays, It is very difficult for doctors to detect a brain tumor at an early stage. MRI images are more susceptible to noise and other environmental disturbances. Therefore, it becomes difficult for doctors to determine whether the tumor is present ornot.

This paper proposes a comprehensive approach, initially employing the k-nearest Neighbors (KNN) classifier and subsequently transitioning to a Convolutional Neural Network (CNN) for enhanced performance. For the KNN classifier, data normalization and reshaping ensure compatibility, followed by training and prediction. However, dueto the limitations of the KNN classifier such as sensitivity to noise and feature complexity, KNN classifier performance is low. So CNN Classifier is used. The CNN architecture integrates convolutional layers for feature extraction, ReLU activation for non-linearity, max-pooling for spatial reduction, flattening for dense layer input, and dense layers for classification. The Adam optimizer is employed to refine the performance of the Convolutional Neural Network (CNN), enhancing its ability to accurately extract features and classify brain tumor images from MRI scans. By usingthe Kaggle dataset accuracy of brain tumor detection using KNN was 75% and by CNNaccuracy was 96%.

Smart Wate Managment: A Deep Dive into Garbage Classfication With Deep Learning

E. Ramesh
Computer Science and Engineering, Annamacharya Institute of Technology and Sciences, Rajampet, eeramesh8@gmail.com
D. Mahesh
Computer Science and Engineering, Annamacharya Institute of Technology and Sciences, Rajampet, maheshalg2001@gmail.com
P. Lokesh
Computer Science and Engineering, Annamacharya Institute of Technology and Sciences, Rajampet, lokeshpappuru00@gmail.com
D. Nikhil
Computer Science and Engineering, Annamacharya Institute of Technology and Sciences, Rajampet, nikhildoppani3@gmail.com
S. Asif
Computer Science and Engineering, Annamacharya Institute of Technology and Sciences, Rajampet, singanamalaasif678@gmail.com

ABSTRACT

Recycling of waste from households and industries, is one of the methods that has been proposed to reduce the ever-increasing pressure on landfills. Different types of waste types warrant different management techniques and hence, proper waste segregation according to its types is essential to facilitate proper recycling. Current existing segregation method still relies on manual hand - pickingprocess. In this paper, a method; based on deep learning is used to classify wastes using their images into six different waste types (glass, metal, paper, plastic, cardboard and trash) has been proposed. The suggested approach's effectiveness and scalability are demonstrated by the experimental findings, underscoring its potential for practical implementation in a range of environmental conservation activities. Here, Convolutional Neural Network (CNN) model has been used for classification of waste.

Knapds: Knowledge Navigator and Ai-Powered Document Summarizer

Kashish Deepak Lalwani
CSE AIML, D. Y. Patil College of Engineering and Technology
Kolhapur, India, kaashishlalwani@gmail.com
Nihal Naushad Sayyad
CSE AIML, D. Y. Patil College of Engineering and Technology
Kolhapur, India, nihal.n.sayyad@gmail.com
Pranav Shridhar Sawant
CSE AIML, D. Y. Patil College of Engineering and Technology
Kolhapur, India, pranavsawant2579@gmail.com
Digvijay Dattatray Patil
CSE AIML, D. Y. Patil College of Engineering and Technology
Kolhapur, India, patildigvijay7878@gmail.com
Avantika Ramprasad Patil
CSE AIML, D. Y. Patil College of Engineering and Technology
Kolhapur, India, avantikapatil940@gmail.com
Dr. Siddheshwar. V. Patil
CSE AIML, D. Y. Patil College of Engineering and Technology
Kolhapur, India, siddheshwar.patil@gmail.com

Abstract

In today's information-driven world, the abundance of textual data presents a challenge for efficient comprehension and analysis. As students, researchers, and professionals alike, we often find ourselves grappling with lengthy PDF documents containing critical information. This paper sets out to address this challenge by harnessing the capabilities of the OpenAI API to automate the summarization of complex PDF texts. The primary objective of this paper is to develop a system that utilizes OpenAI's powerful language model to extract key insights and condense the content of PDF documents into concise summaries. By leveraging the advanced natural language processing (NLP) capabilities of the OpenAI API, the system aims to improve the efficiency of information assimilation and enhance work for users. This paper follows a systematic approach, commencing with data pre-processing to handle the extraction of textual content from PDF files effectively. The extracted data is then passed through the OpenAI API, which employs state-of-the-art deep learning techniques to analyse and comprehend the text. Through iterative refinement and fine-tuning, the application generates coherent and meaningful summaries that capture the essence of the original document. The significance of this work lies in its potential to revolutionize how we consume and utilize information from lengthy documents. By simplifying complex texts into concise summaries, the application can save time and effort for students, researchers, and professionals, allowing them to focus on higher-order tasks and make well-informed decisions based on relevant insights.

Keywords— Large Language Model, OpenAI, Natural Language Processing.

Multispectral Image Super-Resolution Via Multiple-Frame Splicing and Degradation Learning

Samreen Fiza
Department of Electronics and Communication Engineering, School of Engineering, Presidency University, Bangalore, Karnataka, India; samreenfiza236@gmail.com,
Sandhya Tatekalva
Department of Computer Science, S. V. University, Tirupati, Andhra Pradesh, India, dr.sandhyasatish@gmail.com
G. Tirumala Vasu
Department of Electronics and Communication Engineering, School of Engineering, Presidency University, Bangalore, Karnataka, India; sietk81nitt@gmail.com

ABSTRACT

An up-and-coming method for distinguishing between various objects detected by remote sensing is multispectral imagery (MSI). The ensuing object detection applications are limited by the relatively low MSI spatial resolution, which is a result of the trade-off between limited physical hardware and different imaging conditions. Fusion hyperspectral image super resolution (FHSR) requires additional RGB/multispectral data, which isn't practical for most MSI applications, while single multispectral super resolution (SMSR) approaches are currently unable to extract finer details. Neither of the two kinds of existing approaches pays enough attention to the many sources of degradation that result from low spatial resolution. This article presents an advancement in the design of a new framework for multiple frame splicing strategies, which significantly enhances SMSR quality, and shows how to apply several MSI degradation models to better match the actual degradation situation. This framework is an end-to-end super resolution (SR) network, which eliminates the need for a complicated attention residual model and eliminates the need for a single up-sampling module. It does this by using several splicing low-resolution input samples with high-resolution outputs, all of the same size. Faster training convergence and less haziness at higher multiples are two benefits of a well-designed framework. This framework allows for the simultaneous combination of numerous low-resolution degradation samples, which improves the fit for the blind SR result. Deterioration specifically targets blur, noise, compression, and hybrids thereof in order to mimic actual degradation. The experimental findings on three separate hyperspectral datasets show that the suggested algorithm can beat the state-of-the-art SMSR approaches and greatly improve the details in the recovered high-resolution hyperspectral images.

Keywords: Super-Resolution, Multispectral Image, Object detection, Remote sensing

Cabuto Fractional Differential Based Medical Image Segmentation

Samreen Fiza
Department of Electronics and Communication Engineering, School of Engineering, Presidency University, Bangalore, Karnataka, India; samreenfiza236@gmail.com,
Sandhya Tatekalva
Department of Computer Science, S. V. University, Tirupati, Andhra Pradesh, India, dr.sandhyasatish@gmail.com
G. Tirumala Vasu
Department of Electronics and Communication Engineering, School of Engineering, Presidency University, Bangalore, Karnataka, India; sietk81nitt@gmail.com

Abstract

Intelligent medical diagnostics and other cutting-edge medical applications rely heavily on medical image segmentation technology, which has garnered considerable interest. Nevertheless, precise "focus" target segmentation is extremely challenging because to the gray heterogeneity seen in medical images. Fractional calculus's strong nonlinear processing properties have made it a hot topic in the image processing industry in recent years. This article proposes the Caputo CV model, a medical image segmentation algorithm based on Caputo fractional differentia. The model is theoretically shown to be feasible after the introduction of the Caputo fractional differential with weak singularity. On the other hand, the numerical model calculation makes use of the high-order Caputo fractional differential mask operator with an order in the interval (2, 3) to process the fitting term of the model, and the low-order Caputo fractional differential operator with an order in the interval (0, 1) to discretize the gradient in the approximation term, in accordance with the properties of the two types of operators. This allows the model to stabilize and improve its evolution process by improving the high-frequency information while effectively maintaining the nonlinear properties of the image's low-frequency data during iteration. Numerous tests demonstrate that the suggested methodology significantly decreases running time while significantly improving accuracy of gray inhomogeneous target segmentation.

Keywords: medical image segmentation, Cabuto fractional differential, Mask operator, target segmentation

Enabling Effective Equalization and Carrier Frequency Offset Compensation for Wireless Communication Systems Underwater

Samreen Fiza
Department of Electronics and Communication Engineering, School of Engineering, Presidency University, Bangalore, Karnataka, India; samreenfiza236@gmail.com,
Sandhya Tatekalva
Department of Computer Science, S. V. University, Tirupati, Andhra Pradesh, India, dr.sandhyasatish@gmail.com
G. Tirumala Vasu
Department of Electronics and Communication Engineering, School of Engineering, Presidency University, Bangalore, Karnataka, India; sietk81nitt@gmail.com

Abstract

Underwater Acoustic (UWA) wireless communication devices suffer from numerous issues that limit their performance. Reasons for this include the fact that seawater has a large attenuation factor, the kind of sediment, the concentration of acidity, the temperature of the water, and the speed of sound transmission. One such approach is Orthogonal Frequency Division Multiplexing (OFDM). The Inter-Carrier-Interference (ICI) phenomenon, which affects OFDM systems, is a major drawback of these systems. One solution to this challenge is combined equalization and CFO compensation. In the current research, the standard OFDM system is adapted for Multiple-Input-Multiple Output (MIMO)-OFDM communication by using

Discrete Wavelet Transform (DWT) instead of Discrete Fourier Transform (DFT). The DWT-based OFDM system has several advantages over the similar DFT. The trade-off, on the other hand, is the need for an additional DFT/IDFT to complete the Frequency-Domain equalization method, increasing overall computing complexity. In addition, we provide a Joint Low Regularized Linear Zero Forcing equalizer for MIMO-OFDM based on DWT that uses the banded-matrix approximation method. The proposed method foregoes signal-to-noise ratio estimation. Simulation findings show that the proposed scheme outperforms other schemes spatially at the same UWA channel conditions in the presence of estimate errors.

Key words: MIMO, OFDM, CFO, UWA channel.

Complexity Reduced Universal Mrc Detector for Otfs

Afreen Kubra
Department of Electronics and Communication Engineering, School of Engineering, Presidency University, Bangalore, Karnataka, India, afreenkubra96@gmail.com
Samreen Fiza
Department of Electronics and Communication Engineering, School of Engineering, Presidency University, Bangalore, Karnataka, India, samreenfiza236@gmail.com
G. Tirumala Vasu
Department of Electronics and Communication Engineering, School of Engineering, Presidency University, Bangalore, Karnataka, India, sietk81nitt@gmail.com

Abstract

Orthogonal time frequency space (OTFS) is a newly suggested modulation technique that combines information symbols in the delay-Doppler domain to address significant Doppler shifts in wireless communications with high mobility. This paper categorizes all the different variations of OTFS based on whether a cyclic prefix (CP) or zero padding (ZP) is applied to each individual block or to the entire frame as a whole.

Next, we provide the overall input-output connections for both integer and fractional delays and Doppler shifts. In addition, we introduce a universal maximum ratio combining (MRC) detector for all OTFS versions that has the lowest complexity compared to other known OTFS detection systems, without sacrificing performance. Ultimately, we establish the relationship between the input and output for channels with a wide range of delays, where the delays can be longer than the period of the OTFS block. The simulation findings indicate that the MRC detection provides consistent performance across all OTFS variations and is also effective for channels with overspread.

Advanced Visual Cryptography Technique for Secure Transmission of Grayscale Medical Images with Meaningful Progression

G. Tirumala Vasu
Department of Electronics and Communication Engineering, School of Engineering, Presidency University, Bangalore, Karnataka, India, sietk81nitt@gmail.com
Sandhya Tatekalva
Department of Computer Science, S. V. University, Tirupati, Andhra Pradesh, India, dr. sandhyasatish@gmail.com
Samreen Fiza
Department of Electronics and Communication Engineering, School of Engineering, Presidency University, Bangalore, Karnataka, India, samreenfiza236@gmail.com

Abstract

Data transmission is critical in the digital world, and it can take many forms, including text, photos, and videos. Banking, medical, marketing, social media, and commercial apps all rely significantly on visuals to communicate with clients and others. However, the rising reliance on digital communication has attracted Cyber Attackers (CAs), which are individuals or institutions that exploit network gaps and disrupt data integrity, resulting in data transmission vulnerabilities. The healthcare business, in particular, has borne enormous losses as a result of data breaches, according to the Cost of a Data Breach 2021 report. Securing such data during communication presents a significant problem. Visual cryptography

(VC) appears as an effective method for secure picture sharing. The VC's share creation module encrypts the image and divides it into several shares, which are then sent to recipients. Individual shares do not reveal classified information. At the destination, the shares are digitally or physically stacked to rebuild the original image without requiring sophisticated mathematical procedures or additional hardware. Key limits in VC include share picture security, reconstruction quality, pixel expansion difficulties, and computational complexity. Despite the importance of venture capital, existing research publications frequently focus on specific types of schemes, resulting in gaps in complete comparisons and coverage of the state of the art. To overcome this, we present a new Progressive Meaningful Visual Cryptography (PMVC) approach for exchanging secret images that causes an error instance. The suggested method delivers meaningful shares while maintaining image quality, with a peak signal-to-noise ratio (PSNR) of up to 37 dB and no pixel expansion concerns. Furthermore, the proposed approach works with (n, n) VC schemes and accepts grayscale secret photos.

Key words: Visual Cryptography, Medical images, Single grayscale secret, Meaningful shares.

Reliable and Low Papr System for Ofdm Using TCCM

G. Tirumala Vasu
Department of Electronics and Communication Engineering, School of Engineering, Presidency University, Bangalore, Karnataka, India, sietk81nitt@gmail.com
Sandhya Tatekalva
Department of Computer Science, S. V. University, Tirupati, Andhra Pradesh, India, dr. sandhyasatish@gmail.com
Samreen Fiza
Department of Electronics and Communication Engineering, School of Engineering, Presidency University, Bangalore, Karnataka, India, samreenfiza236@gmail.com

Abstract

To convey multimedia data, rapid communication makes use of orthogonal frequency division multiplexing (OFDM). OFDM reduces frequency-selective fading and improves bandwidth efficiency. With multiple subcarriers, OFDM has a high peak-to-average power ratio (PAPR), which limits its potential. The transmitted waveform's nonlinearity is due to excessive PAPR. Because the OFDM signal is sparse in the frequency domain, this work uses compressed sensing (CS) to reduce PAPR. Thus, the transmitter multiplies a well-designed topologically conjugate chaotic circulant matrix (TCCM), while the receiver employs orthogonal matching pursuit. The TCCM involves a large number of topologically conjugate chaotic functions. Chaotic matrices are preferred because they ensure secure data transfer, with any little change in the chaotic parameters resulting in irrecoverable data. The proposed chaotic system is evaluated with the bifurcation diagram (BD), Lyapunov exponent (LE), and so forth. This structured matrix significantly reduces the PAPR from 13 dB to less than 7. 5 dB, and its assessment measure is the CCDF (complementary cumulative distribution function). The OFDM system's picture transmission is compared using a Gaussian matrix (GM), resulting in better PSNR, BER, and lower PAPR. This approach encrypts data while lowering the PAPR, making it appropriate for all future networks, including cognitive and 5G networks.

Key words: PAPR, OFDM, PSNR, Compressed sensing, BER.

Design & Implementation of 8th Sram Cell Using 22nm for Low Power Applications

Giddaluru Harshavardhan
M. Tech (VLSI-SD), Srinivasa Institute of Technology and Science, Kadapa.
Mr. R. L. B. R Prasad Reddy
Associate Professor, Department of E. C. E., Srinivasa Institute of Technology and Science, Kadapa.

ABSTARCT

Static Random Access Memory (SRAM) plays a vital role as a component of physical memory in computing systems, offering high speed and volatility. However, its inherent power-hungry nature poses challenges in modern low-power applications. This research presents a novel approach to address this issue through the design and implementation of an 8T SRAM cell using advanced 22nm semiconductor technology optimized for low-power applications. The proposed 8T SRAM cell demonstrates significant improvements in both read and write operations compared to conventional Semi Adiabatic and Quasi-Adiabatic designs. Through meticulous optimization, the proposed SRAM cell achieves enhanced performance while minimizing delay and power consumption, thereby contributing to the overall efficiency of the computing system. This paper underscores the importance of innovative circuit design techniques in mitigating power consumption challenges in modern semiconductor technologies, paving the way for energy-efficient computing solutions in diverse application domains.

Keywords: SRAM, Power consumption, Adiabatic, VLSI, 8T SRAM cell, Quasi-Adiabatic, 22nm delay.

Controlling Power Point Presentation Wi-fi Enabled Gloves Using Hand Gestures

P. Voshanavi
UG Students, Department of ECE,
GMR Institute of Technology, Rajam
S. Vamsi
UG Students, Department of ECE,
GMR Institute of Technology, Rajam
P. Sri Sai Nandu
UG Students, Department of ECE,
GMR Institute of Technology, Rajam
P. Vinod
UG Students, Department of ECE,
GMR Institute of Technology, Rajam
B. Anil Babu
Professor, Department of ECE,
GMR Institute of Technology, Rajam
Corresponding Authors:
21341A04D9@gmrit.edu.in, anilbabu.b@gmrit.edu.in

ABSTRACT

This paper introduces a novel method for enhancing PowerPoint presentations by enabling control through hand gestures. Traditional remote controls and accessories often limit mobility and audience engagement, prompting the need for a more interactive solution. This paper is specially designed gloves equipped with sensors and wireless communication modules, such as Wi-Fi or Arduino microcontrollers. These gloves incorporate accelerometer/gyroscope modules to detect various hand gestures accurately. By seamlessly integrating gesture recognition technology, presenters can navigate slides and execute commands with fluid hand movements. This innovation not only enhances presenter mobility but also fosters increased audience involvement and attention. The main objective of this model is to revolutionize the presentation experience, making it more dynamic and engaging for both presenters and audiences. The paper aims to bring presenters and audiences closer together, making presentations more exciting and involving. The hand gestures are used to control slides, which makes presentations more interactive.

KEYWORDS: Power Point presentations, Hand gestures, Gesture control, Arduino microcontroller, Wi-Fi module, Accelerometer, Interactive presentations.

Arduino Based Electronic Voting System Using Fingerprint Sensor

P. Sravani
UG Students, Department of ECE,
GMR Institute of Technology, Rajam
P. Kousik
UG Students, Department of ECE,
GMR Institute of Technology, Rajam
Adityaraj
UG Students, Department of ECE,
GMR Institute of Technology, Rajam
sK N Jakeer
UG Students, Department of ECE,
GMR Institute of Technology, Rajam
P. RaviKumar
Assistant Professor, Department of ECE,
GMR Institute of Technology, Rajam
Corresponding Authors:
21341A04D7@gmrit.edu.in; Ravikumar.p@gmrit.edu.in

ABSTRACT

Ensuring the integrity and security of electronic voting systems is paramount in modern democracies. This paper presents a novel approach to electronic voting by introducing an Arduino-based system integrated with fingerprint authentication. Leveraging Arduino's decentralized block chain platform, the proposed system guarantees transparency, immutability, and resistance to tampering. By incorporating a fingerprint sensor, the system adds an extra layer of security, preventing identity fraud and unauthorized access. This abstract provides an overview of the proposed Arduino-based electronic voting system, emphasizing its innovative use of biometric authentication to bolster trust and confidence in the electoral process. The integration of fingerprint authentication not only enhances security but also streamlines the voting process, ensuring that each vote is cast by a verified individual. Overall, this abstract highlights the potential of Arduino-based electronic voting systems to advance electoral integrity in the digital age.

KEYWORDS: Arduino uno, I2C lcd display, R307 fingerprint module, speaker, push buttons.

Arduino Based Ultrasonic Smart Glasses for Visually Impaired

V. Charishma
UG Students, Department of ECE,
GMR Institute of Technology, Rajam
T. Murali
UG Students, Department of ECE,
GMR Institute of Technology, Rajam
V. Manasa Aparna
UG Students, Department of ECE,
GMR Institute of Technology, Rajam
T. Gopalakrishna
UG Students, Department of ECE,
GMR Institute of Technology, Rajam
Yogesh Mishra
Professor, Department of ECE,
GMR Institute of Technology, Rajam
Corresponding Authors
21341A04I6@gmrit.edu.in; yogesh.m@gmrit.edu.in

ABSTRACT

Arduino based Ultrasonic smart glasses designed to improve the wayfinding and spatial awareness of blind people. The suggested system combines sophisticated signal processing methods, ultrasonic sensors, and an intuitive user interface to deliver environmental feedback in real-time, allowing users to move through their surroundings more independently and safely. Using ultrasonic waves, the Ultrasonic Smart Glasses identify objects and obstacles in the user's immediate vicinity. The ultrasonic sensors are positioned thoughtfully on the frame of the glasses to record the surrounding environment. The data is then processed through an algorithmic process to ascertain the precise location and distance of obstacles. By converting this data into intuitive feedback, like haptic feedback or auditory cues, the system enables users to sense their environment and make wise navigational decisions. It features an integrated sensor that, at most, scans a range of 30 degrees, distributing ultrasonic waves in the direction the user is travelling. The sensor notices the obstruction as soon as it appears and transmits the information to the gadget that plays an automated speech through an earphone that is inserted into the user's ear.

KEYWORDS: Ultrasonic Sensors, Navigation Aid, Arduino, Auditory Cues, Smart Glasses.

Underwater Image Enhncement

B. V. Sowjanya
Department Of Electronics And Communication Engineering,
JNTUA College Of Engineering Pulivendula
G. Kalyani
Department Of Electronics And Communication Engineering,
JNTUA College Of Engineering Pulivendula
N. Nikitha Reddy
Department Of Electronics And Communication Engineering,
JNTUA College Of Engineering Pulivendula
G. Harshavardhan
Department Of Electronics And Communication Engineering,
JNTUA College Of Engineering Pulivendula
SK. Hanifa
Department Of Electronics And Communication Engineering,
JNTUA College Of Engineering Pulivendula

Abstract

Underwater images find application in various fields, like marine research, inspection of aquatic habitat, underwater surveillance, identification of minerals, and more. However, underwater shots are affected a lot during the acquisition process due to the absorption and scattering of light. As depth increases, longer wavelengths get absorbed by water; therefore, the images appear predominantly bluish-green, and red gets absorbed due to higher wavelength. These phenomenon's result in significant degradation of images due to which images have low contrast, color distortion, and low visibility. Hence, underwater images need enhancement to improve the quality of images to be used for various applications while preserving the valuable information contained in them. This paper proposes comprehensive approach to enhance the underwater images, by using methods such as color correction, contrast enhancement, and image sharpening to enhance the images. To obtain the enhanced images, we use Gray world algorithm, Histogram equalization and Unsharp masking algorithms

Enhancing Crop Disease Detection With Convolutional Neural Networks Through Transfer Learning and Data Augmentation

Dr. M. Subbarao
Professor of CSE, Dept of CSE, Annamacharya Institute of Technology & Sciences, New Boyanapalli, Rajampet, Boyanapalli, Rajampet – 516115, subbarao@gmail.com
Hyndavi Rachamallu
Dharani Murari, Gnana Harika Lachayagari, Ashreen Shaik
Dept of Computer Science Engineering, Annamacharya Institute of Technology and Sciences, New Boyanapalli, Rajampet, AP – 516126

Abstract

Crop disease detection is a critical aspect of modern agriculture, enabling early intervention and minimizing crop yield losses. Convolutional Neural Networks (CNNs) have shown promise in automating this task by analyzing images of crop leaves or plants. This research focuses on enhancing the accuracy and robustness of CNN-based crop disease detection using two key techniques: Transfer Learning and Data Augmentation. Transfer Learning entails utilizing readily trained CNN models for extracting significant insights from crop images and adjusting them for the purpose of disease classification. This approach mitigates the need for large annotated datasets, making it suitable for resource-constrained agricultural settings. Data Augmentation is another crucial aspect addressed in this study. It entails artificially enlarging the training dataset by implementing diverse alterations (such as rotating, mirroring, and resizing) on the original images. Data augmentation aids in preventing overfitting, especially when training data is limited, and enhances the model capacity for adapting to a variety of environmental conditions and disease presentations. However, several challenges persist in this domain. Ensuring the transferability of features learned from pre-trained models to specific crop diseases and varieties remains a concern. Moreover, fine-tuning parameters and selecting appropriate augmentation strategies require careful consideration. Additionally, addressing class imbalance, handling noisy or unstructured data, and optimizing the model's computational efficiency are ongoing challenges. This study seeks to tackle these concerns and push the boundaries of CNN-based crop disease detection, ultimately making a valuable contribution to fostering sustainable and productive agriculture.

Sales Insight in Real Time with Mysql Integration and Alert System

Dhivya D
Department of Artificial Intelligence and Data science,
Vel Tech Multitech Dr. Rangarajan Dr. Sakunthala Engineering College, Chennai. India.
Nithyashree M
Department of Artificial Intelligence and Data science,
Vel Tech Multitech Dr Rangarajan Dr. Sakunthala Engineering College, Chennai. India.
Abimanya S
Department of Artificial Intelligence and Data science,
Vel Tech Multitech Dr. Rangarajan Dr. Sakunthala Engineering College, Chennai. India.
Sathya D
Department of Artificial Intelligence and Data science,
Vel Tech Multitech Dr. Rangarajan Dr. Sakunthala Engineering College, Chennai. India.

Abstract

"Sales Insight In Real Time With MySQL Integration And Alert System" is a paper that introduces a innovative technique to provide real-time sales data and low stock notifications for the e-commerce business. It achieves this by establishing a live connection between the website and a MySQL database using Python. This Python script acts as a bridge, constantly pushing sales data from your website to the MySQL database. Our system uses Power BI's MySQL connectivity to create dynamic dashboards that let stakeholders rapidly track sales performance. Through frequent updates from

MySQL databases, the system makes sure insights are based on the latest data. Additionally, our solution includes sophisticated alerting mechanisms to inform users of changes in inventory levels. By leveraging Power BI's features, customers can make proactive decisions by being notified in real time when stocks rise or fall below predetermined criteria. In addition to providing organizations with actionable data, this integrated strategy makes it easier to respond quickly to changes in the market. Our research shows how real-time sales analytics may be used to maximize revenue potential and optimize operational efficiency.

Smart Guarding Women Safety Enhanced With Iot and Sensor Networks

C. Sashank Reddy
Department of Electronics and Communication Engineering,
UG STUDENT, Annamacharya University, Rajampet – 516126
N. Sreenath Reddy
Department of Electronics and Communication Engineering,
UG STUDENT, Annamacharya University, Rajampet-516216
G. V. Jeevan Kumar Reddy
Department of Electronics and Communication Engineering,
UG STUDENT, Annamacharya University, Rajampet - 516126
M. Venkata Naresh
Department of Electronics and Communication Engineering,
UG STUDENT, Annamacharya University, Rajampet - 516126

Abstract

In spite of modern advancements in technology, women's safety continues to be a major concern. Women are particularly vulnerable when traveling alone on lonely roads or in deserted areas. Current hand-held safety devices require the user to manually activate them by pressing a button or fingerprint authentication and automatically alerts family persons and the police when a woman is in danger. We propose a solution that addresses these issues and provides reliable safety for women.

If the device detects an unsafe situation it will send out an alert. Additionally, the device includes a shockwave generator that women can use to defend themselves against attackers. The proposed design also includes additional features such as the location also gets updated in the things speak server when the device gets activated .

Overall, our solution aims to provide women with a high level of security and peace of mind.

Predictive Analytics and Intervention Strategies for Enhanced Road Safety: a Machine Learning Approach

Dr. Penchalaiah . N
Associate Professor, Dept. of CSEAITS, Rajampet
Rajampet, AP, India penchalaiah550@gmail.com
Haritha. Y
Dept. of CSE AITS, Rajampet, AP, India
harithayerramaneni@gmail.com
Chandana. G
Dept. of CSE AITS, Rajampet, AP, India
chandana868893@gmail.com
Gowthami. T
Dept. of CSE AITS, Rajampet, AP, India
thupakulagowthami@gmail.com
Geetha Pavani. Y
Dept. of CSE AITS, Rajampet, AP, India
chinna791711@gmail.com

Abstract

The increasing frequency and severity of global road accidents necessitate novel approaches to bolster road safety. This study addresses this imperative by suggesting a predictive analytics framework powered by machine learning to anticipate and prevent road accidents. Despite advancements in road safety measures, the persistence of accidents demands a proactive and data-driven paradigm. Utilizing machine learning algorithms, this research aims to scrutinize various data sources, encompassing traffic patterns, weather conditions, and driver behaviour, in order to construct a robust predictive model. Nevertheless, this undertaking is not devoid of challenges. Moreover, ensuring the privacy and ethical use of personal data in the smart transportation systems era remains a paramount consideration. This research endeavors to surmount these challenges through the development of clear and understandable machine learning models, ethical data handling practices, and collaboration with pertinent stakeholders. By proposing inventive intervention strategies rooted in predictive analytics, this study aspires to contribute to the continuous initiatives aimed at establishing safer and more sustainable road networks.

Square Patch Antenna with Slots and DGS for 5g and LTE Applications

S. Kavya
Department of ECE, JNTU College of Engineering
Pulivendula, AP, India, singareddykavya123@gmail.com
K. Divya
Department of ECE, JNTU College of Engineering
Pulivendula, AP, India, kalimisettydivya@gmail.com
B. Goutham
Department of ECE, JNTU College of Engineering
Pulivendula, AP, India, bgowtham2002@gmail.com
V. Prasanna Kumar
Department of ECE, JNTU College of Engineering
Pulivendula, AP, India, vaddiprasannakumar359@gmail.com
P. Mohith Kumar Achari
Department of ECE, JNTU College of Engineering
Pulivendula, AP, India, pirangimohithkumar@gmail.com
Dr. R. Ramana Reddy
Professor, Department of ECE, JNTU College of Engineering
Pulivendula, AP, India, profrrreddy@yahoo.co.in

Abstract

This paper introduces a square patch antenna with slots and DGS for 5G and LTE applications, featuring pentagon and hexagon slots at the centre of the square patch. The antenna's dimensions are 37 × 34 × 1. 6 mm3, with the patch measuring 17 × 17 mm2. The square patch antenna with pentagonal and hexagonal slots was designed. The hexagonal slotted patch antenna yielded a bandwidth of 110MHz with a peak gain of 5dBi. The pentagonal slotted patch antenna resonates at 3. 95GHz with a bandwidth of 120MHz (3. 88GHz to 4GHz) and a peak gain of 5dBi. By incorporating DGS, the bandwidth increased to 2. 08GHz and the peak gain to 5. 4dBi, with a resonating frequency of 4. 15GHz. Simulations were carried out using Ansys HFSS software.

Index Terms- Square Patch (SP), Slots, Defective Ground Structure (DGS).

Enhancing Intrusion Detection System Performance Through Principal Component Analysis and Machine Learning Classifiers: a Comprehensive Investigation

P. Swathi
Assistant Professor, Dept. of CSE AITS, Rajampet, AP, India
1260swathi@gmail.com
Bhavya Sree. K
Dept. of CSE AITS, Rajampet, AP, India
kbhavyasreevarma@gmail.com
Arshiya. S
Dept. of CSE AITS, Rajampet Rajampet, AP, India
shaikarshiya742@gmail.com
Bindu Priya. G
Dept. of CSE AITS, Rajampet Rajampet, AP, India
gowrambindupriya@gmail.com

Abstract

In the swiftly changing digital environment of today, the escalating sophistication of cyber threats presents a considerable hurdle to the efficiency of Intrusion Detection Systems (IDS). This study addresses the urgent requirement to improve IDS performance by adopting an innovative approach that combines Principal Component Analysis (PCA) with cutting-edge machine learning classifiers. The research delves into a thorough examination of incorporating PCA as a technique for feature reduction, aiming to mitigate the challenges associated with the curse of dimensionality and consequently enhance the efficiency and computational economy of IDS. Current challenges encountered by contemporary IDS encompass the accurate identification of varied and intricate intrusion patterns while minimizing false positives. This investigation delves into the complex interconnections within network data through PCA, extracting pivotal features that contribute to detecting subtle anomalies. Subsequently, machine learning classifiers like support vector machines, random forests, and neural networks are applied to scrutinize the condensed feature space, augmenting the system's capability to recognize both familiar and unprecedented intrusions. Through a comprehensive exploration of the amalgamation of PCA and machine learning classifiers, this study strives to present a resilient and scalable solution to the prevailing challenges in intrusion detection.

Keywords---Intrusion Detection Systems, Principal Component

Analysis (Pca), Machine Learning Classifiers, Cyber Security, Curse of Dimensionality, Network Anomalies

DIABETIC RETINOPATHY USING CNN

Shaik Taj Mehboob
Associate Professor and HOD, Department of ECE,
JNTUA College of Engineering Pulivendula.
O. Glysia
Student, Department of ECE,
JNTUA College of Engineering Pulivendula.
K. Devendra
Student, Department of ECE,
JNTUA College of Engineering Pulivendula.
S. Noorulla
Student, Department of ECE,
JNTUA College of Engineering Pulivendula.
B. Hemanth
Student, Department of ECE,
JNTUA College of Engineering Pulivendula.
S. Noor Basha
Student, Department of ECE,
JNTUA College of Engineering Pulivendula

ABSTRACT

Diabetic Retinopathy (DR) is a common complication of diabetes that can lead to visual impairment and blindness if not managed effectively. In the present scenario, diabetic retinopathy remains a significant concern globally, with millions of individuals at risk of developing this condition. Early diagnosis is crucial to prevent vision loss. Traditional methods of identifying DR require manual examination by ophthalmologists, which can be time-consuming and prone to human error.

This paper proposes the use of automated detection of DR model from retinal images. And also explains how the model shows difference in prediction with various sets of training and testing datasets. The model got an overall accuracy of 97% when four cases are considered where training images are 8%, 12%, 40%, and 50% and testing images are 92%, 88%, 60%, and 50% respectively. Once the model is trained and saved, it can predict whether the input fundus image is having the Diabetic Retinopathy or not. If the fundus image contains the DR it classifies the severeness of Diabetic Retinopathy.

KEYWORDS: Diabetic Retinopathy, Fundus image, CNN, Deep Learning, Image classification

Ecg Signal Extraction Method Based on Singular Value Selection and Wavelet

E. SINDHU
Department of ECE, Srinivasa Institute Of Technology And Science, Kadapa, Mail id:-sindhuindla143@gmail.com
J. ASHOK REDDY
Department of ECE, Srinivasa Institute Of Technology And Science, Kadapa, Mail id:-jonnavaramashokreddy@gmail.com

Abstract

Aiming at the noise of the received ECG signal, a extraction method of ECG signal based on singular value selection and wavelet is proposed. The singular value decomposition on the ECG signal is performed firstly, and the ECG signal component corresponding to each singular value is obtained. Then the signal component corresponding to the maximum singular value is used to calculate cross-correlation coefficients with other components. The cumulative contribution rate of singular values is combined to determine the number of singular values for ECG signal reconstruction. The wavelet threshold de-noising method is used to de-noise the final determined signal components. Finally, the de-noising ECG signal is obtained by reconstructing the signal components. The experimental results show that the method can suppress noise and extract signal effectively, and it has a good noise reduction effect compared with wavelet threshold method.

Keywords- ECG signal, signal extraction, singular value decomposition, wavelet

Content Based Recommendation Systems

Nikitha. P
Assistant Professor, Dept of CSE, AITS, Rajampet.
T. Charpitha Sree
Dept of CSE, AITS, Rajampet AP, India.
Bramhani. Y
Dept of CSE, AITS, Rajampet AP, India.
Asritha. S
Dept of CSE, AITS, Rajampet AP, India.
Ayyavaramma. M
Dept of CSE, AITS, Rajampet AP, India.

Abstract

The amount of information in e-commerce is increasing far more quickly than our ability to process. Recommender systems apply knowledge discovery techniques to help people find what they really want. However, all of the previous approaches have an important drawback: items added newly cannot be found. In this paper, a general framework is proposed for supporting automatic recommendation of the new item to the potential users based on the concept of influent sets. We propose a simple efficient indexing structure and a heuristic information retrieval technique algorithm for searching reverse k nearest neighbour in high-dimensional dataset. And experimental evaluation reveals that our approach outperforms the previous algorithm and enhances the performance efficiently.

Enhanced Thermal Management in 3d Integrated Circuits

Surla Neelima
Student, Dept of ECE, JNTUACEP, AP, India,
surlaneelima2002@gmail.com
Palla Dileep
Student, Dept of ECE, JNTUACEP, AP, India,
dileeppalla18@gmail.com
Talachandra Avinash
Student, Dept of ECE, JNTUACEP, AP, India,
talachandraavinash@gmail.com
Galla Vandana
Student, Dept of ECE, JNTUACEP, AP, India,
vandanareddy496@gmail.com
Challa Kambagiri Raju
Student, Dept of ECE, JNTUACEP, AP, India,
rajuramudu111@gmail.com
M. Siva Kumar
Assistant Professor, Department of ECE, JNTUACEP, AP, India,
sivakumar. ece@jntua. ac. in

Abstract

Three Dimensional IC (3D IC) is a new technology that may successfully overcome the obstacles associated with 2D IC integration. Thermal difficulties among layers and noise coupling among TSV tosubstrate and TSV-to-TSV are important limitations in 3D IC architectures, which have a considerable impact on total system performance. In 3D ICs, the applied potential increases the heat of the IC, but the problem stated is how the impact during the heat on the direction heat travels to the heat sink, heat travelling to the heat source to the heat sink in a specific direction, this component will be damaged and there is a chance of the IC failing. In this work presents a FIN with optimal orientations and heat spreaders using COSMOL multi-physics software that efficiently diffuse heat in different directions and absorb heat towards a heat sink. We also demonstrated various thermal cooling effects and their influence on the distribution of potential throughout the IC and we introduced different dielectric materials such as Al2o3, Si3N4, and Si02 and observed the results based on heat management.

Key Words : 3D IC, Thermal through silicon via (TTSV), Noise Coupling,, CNT, FIN

Detection of Infectious Diseases Using Random Forest

Dr . M. Subba Rao
Professor and Head of Dept. of CSE, Annamacharya Institute of Technology & Sciences, New Boyanapalli, Rajampet – 516115
msraoswap@gmail.com
K. Sai Navya Sree
B. Tech Students, Dept. of CSE, Annamacharya Institute of Technology & Sciences, New Boyanapalli, Rajampet – 516115
sainavyasree. k2003@gmail.com
R. Rani
B. Tech Students, Dept. of CSE, Annamacharya Institute of Technology & Sciences, New Boyanapalli, Rajampet – 516115
roddarani03@gmail.com
S. Sadia Halima
B. Tech Students, Dept. of CSE, Annamacharya Institute of Technology & Sciences, New Boyanapalli, Rajampet – 516115
sadia. shaik2702@gmail.com
D. Sathwika
B. Tech Students, Dept. of CSE, Annamacharya Institute of Technology & Sciences, New Boyanapalli, Rajampet – 516115
sathwikadiyyala2003@gmail.com

Abstract

Viral diseases continue to pose a global health challenge, prompting the need for innovative approaches in prediction, control, and management. Machine learning, recognized for its capacity to derive insights from data, has emerged as a potent tool in this pursuit. Nevertheless, this rapidly evolving field encounters a spectrum of intricate challenges that warrant immediate attention. This research is committed to addressing four pivotal facets: Data

Quality, Generalization, Real-time Adaptation, and Ethical Considerations. To begin with, the process of collecting data related to viral diseases is often marred by inaccuracies, gaps, and biases. We explore recent advancements in data acquisition methods and strategies for improving data quality to mitigate these issues. Moreover, while machine learning models exhibit impressive performance under controlled conditions, their ability to generalize across diverse populations, healthcare systems, and the evolving landscape of viral strains presents a formidable challenge. We delve into innovative techniques to enhance model transferability, ensuring their resilience. Additionally, the rapid mutations and dynamic nature of viral diseases call for real-time adaptability in machine-learning models. Our research focuses on developing algorithms capable of prompt detection and response to emerging viral threats. Lastly, we scrutinize the ethical ramifications of employing machine learning in viral disease management, covering aspects such as patient privacy, fairness, transparency, and consent. We put forth ethical frameworks and governance structures to steer responsible implementation. This research strives to furnish a comprehensive grasp of these critical issues and propose pragmatic solutions to augment the role of machine learning in viral disease management, ultimately contributing to more efficacious, adaptable, and ethically sound interventions in public health.

Keywords: Viral diseases, Machine learning, Data quality, Generalization, Real-time adaptivity, Ethical considerations.

Pneumonia Detection Using CNN Algorithm

V. Nagoorbee
Student, Dept. of ECE, JNTUACEP, AP, India,
venuthurlanagoorbee049@gmail.com
G. Sai Chaithanya
Student, Dept. of ECE, JNTUACEP, AP, India,
saichaitanya1214@gmail.com
P. Usha Sree
Student, Dept. of ECE, JNTUACEP, AP, India,
poolaushasree. royal@gmail.com
Y. Abhishek Reddy,
Student, Dept. of ECE, JNTUCEP, AP, India,
abishekreddyas286@gmail.com
S. Varalakshmi
Assistant Professor, Dept. of ECE, JNTUACEP, AP, India,
varalakshmi1190@gmail.com

Abstract

Pneumonia is form of a respiratory infection that affects the lungs. In these acute respiratory diseases, human lungs which are made up of small sacs called alveoli which is air in normal and healthy people, but in pneumonia these alveoli get filled with fluid or pus. Chest X-rays are used for recognizing the pneumonia infection. Accordingly, to perceive the pneumonia radiologists needs to see the Chest X-rays and he/she wants to revive the expert precisely.

The essential objective of this paper is to recognize if the individual has pneumonia or not with high accuracy using Convolutional Neural Network. Performance of a CNN model over the testing data set is calculated after the completion of the training of the model and the performance is calculated in the following six metrics: accuracy, precision, sensitivity or recall, F1 score, the area under the curve (AUC), and specificity.

Multi-Classification Cnn Model for Brain Tumor Mr Images by Using Resnet-50

Murali Krishna Atmakuri
Department of Electronics and Instrumentation Engineering, Faculty of Engineering and Technology, Annamalai University, Annamalai Nagar, India, muralikrishna. atmakuri80@gmail.com
Ganesh Ram A
Department of Instrumentation Engineeringline, Madras Institute of Technology, Anna University, Chennai, India, agram72@gmail.com
V. V. K. D. V. Prasad
Department of Electronics and Communication Engineering Gudlavalleru Engineering College, Gudlavalleru, India, varrevkdvp@rediffmail.com

Abstract

This paper explores the application of deep learning techniques for predicting the phase of brain tumors, aiming to assist practitioners in assessing patients' conditions and forecasting potential anomalies. Leveraging deep learning facilitates correlating patient data with similar cases, offering a valuable tool for remote healthcare, particularly pertinent during the ongoing pandemic. Digital image digitization and intelligent analysis expedite initial impressions, streamlining manual computations and cross verifications. Deep learning's efficacy in supervised classification of intricate patterns is noteworthy, allowing for the development of fully automated methods using existing toolboxes in simulation environments like Matlab. The study focuses on employing this machine learning approach to classify brain images depicting various tumor types: meningioma, glioma, and pituitary. Evaluation utilizing standard metrics within the Matlab environment reveals confusion matrices indicating accuracy ranging from 89% to 95%. Comparative analysis with Convolutional Neural Network (CNN) and Artificial Neural Network (ANN) models further underscores the effectiveness of the proposed deep learning method.

Implementation of Novel System to Detect Lpg Gas Leakage and fire Using Arduino

V. Rajeswari
UG Students, Department of ECE, GMR Institute of Technology, Rajam, Andhra Pradesh
V. Abhiram
UG Students, Department of ECE, GMR Institute of Technology, Rajam, Andhra Pradesh
V. Chaitanya Sai kumar
UG Students, Department of ECE, GMR Institute of Technology, Rajam, Andhra Pradesh
N. padmavathi
UG Students, Department of ECE, GMR Institute of Technology, Rajam, Andhra Pradesh
Ch. Babji Prasad
Department of ECE, GMR Institute of Technology, Rajam, Andhra Pradesh.
Corresponding Authors:
21341A04I2@gmrit.edu.in; babjiprasad.ch@gmrit.edu.in

ABSTRACT

Gas leakage and fire incidents pose significant risks to life and property. Rapid detection and alert systems are crucial for minimizing potential damages. In this paper, the Gas Leakage and Fire Detection System utilizing Arduino microcontroller and Liquid Crystal Display (LCD), and a buzzer is proposed. The system employs gas and fire sensors to detect hazardous situations. When either gas leakage or fire is detected, the sensors trigger the Arduino microcontroller to activate the alarm mechanism. The LCD screen displays the type of hazard detected, enabling users to take appropriate actions promptly. Additionally, the buzzer alerts nearby residents with an audible sound. This system provides real-time monitoring and alerting, enhancing safety measures in residential, commercial, and industrial settings.

KEYWORDS: Gas leakage detection, Fire detection, Arduino UNO, Home automation, Safety alert

Enhanced Thermal Management in 3d Integrated Circuits

Surla Neelima
Student, Dept of ECE, JNTUACEP, AP, India,
surlaneelima2002@gmail.com
Palla Dileep
Student, Dept of ECE, JNTUACEP, AP, India,
dileeppalla18@gmail.com
Talachandra Avinash
Student, Dept of ECE, JNTUACEP, AP, India,
talachandraavinash@gmail.com
Galla Vandana
Student, Dept of ECE, JNTUACEP, AP, India,
vandanareddy496@gmail.com
Challa Kambagiri Raju
Student, Dept of ECE, JNTUACEP, AP, India,
rajuramudu111@gmail.com
M. Siva Kumar
Assistant Professor, Department of ECE, JNTUACEP, AP, India,
sivakumar. ece@jntua. ac. in

Abstract

Three Dimensional IC (3D IC) is a new technology that may successfully overcome the obstacles associated with 2D IC integration. Thermal difficulties among layers and noise coupling among TSV tosubstrate and TSV-to-TSV are important limitations in 3D IC architectures, which have a considerable impact on total system performance. In 3D ICs, the applied potential increases the heat of the IC, but the problem stated is how the impact during the heat on the direction heat travels to the heat sink, heat travelling to the heat source to the heat sink in a specific direction, this component will be damaged and there is a chance of the IC failing.

In this work presents a FIN with optimal orientations and heat spreaders using COSMOL multi-physics software that efficiently diffuse heat in different directions and absorb heat towards a heat sink. We also demonstrated various thermal cooling effects and their influence on the distribution of potential throughout the IC and we introduced different dielectric materials such as Al2o3, Si3N4, and Si02 and observed the results based on heat management.

Key Words : 3D IC, Thermal through silicon via (TTSV), Noise Coupling,, CNT, FIN

Design of Data Fetcher Decoder for Sram Using GDI Technique

C. V Sai Sudheer
PG Student, Department of ECE, SITS, Kadapa, A. P, India.
D. Kiran Kumar
Asst. Prof, Department of ECE, SITS, Kadapa, A. P, India
RLBR Prasad Reddy
Asso. Prof, Department of ECE, SITS, Kadapa, A. P, India.

Abstract:

In this paper a new method of design of CMOS decoder circuit was proposed. To minimize power consumption and increase the performance of the design the following design solutions have been Designed The New Technique used Is Gate Diffusion Input (GDI) which need Two transistor for design of Basic gates like AND, OR gates which are used in Decoders . In this paper the GDI 2- 4 and 4-16 Decoder is designed and Compared with the Existing Design which is mixed-logic design method, combining transmission gate logic, dual-value logic pass-gate transistor . Three schematics are presented in the paper according which 2-4 decoders were designed. In the first solution a 12- FET based circuit with GDI and 14-FET based LP Decoder and 15-FET Hp Decoders was designed which can be used to minimize the area and power consumption of digital VLSI ICs. The 12- FET based circuit was also proposed to increase the performance of digital ICs with reference to 14-FET and 15-FET based solution. Using the above-mentioned cells, the following more complex digital decoders have been designed: like 4-16 decoders. The above GDI proposed solutions have shown better performances, lower power consumptions and less required area of silicon compared to the ones currently used in digital designs. The proposed solutions have been analyzed in TANNER EDA Software.

Classifying Brain Diseases and Estimatingbrain Age Utilizing Convolutional Neural Network

R. V. Sai Deepthi
Department Of Electronics and Communication Engineering,
JNTUA College Of Engineering Pulivendula
S. P. Farahath Kowsar
Department Of Electronics and Communication Engineering,
JNTUA College Of Engineering Pulivendula
P. Jyoshna
Department Of Electronics and Communication Engineering,
JNTUA College Of Engineering Pulivendula
V. Vinayak Naik
Department Of Electronics and Communication Engineering,
JNTUA College Of Engineering Pulivendula
Miss. N. Shehanaz
Department Of Electronics and Communication Engineering,
JNTUA College Of Engineering Pulivendula

Abstract

The classification of brain diseases and the estimation of brain age using convolutional neural networks (CNN). Brain morphometric pattern analysis has gained traction for identifying age-related imaging biomarkers from structural Magnetic Resonance Imaging (MRI). By analysing MRI images, we classified conditions such as Mild Cognitive Impairment (MCI), Alzheimer's Disease (AD), and Healthy Control (HC). The performance of the traditional Support Vector Machine (SVM) method was assessed, implemented, and compared with the novel CNN approach using Matlab. Our findings indicate that CNNs provide superior accuracy in classifying brain abnormalities compared to SVMs. Additionally, after the classification process, we estimated the range of brain age based on the type of detected abnormality or normality. This dual approach not only enhances the accuracy of brain disease detection but also offers valuable insights into the aging patterns associated with different brain conditions.

This paper proposes a comprehensive approach for brain disease classification and brain age estimation using convolutional neural networks (CNN). By leveraging brain morphometric pattern analysis from structural Magnetic Resonance Imaging (MRI), we effectively identified age-related imaging biomarkers. Our methodology involved classifying brain conditions such as Mild Cognitive Impairment (MCI), Alzheimer's Disease (AD), and Healthy Control (HC) using MRI images. We compared the performance of the traditional Support Vector Machine (SVM) method with the novel CNN approach, demonstrating that CNNs significantly improve classification accuracy.

Smart Sight Extreme Learning Machine (Elm) for Diabetic Retinopathy Detection

Dr. Ch. Nagaraju
Professor & Head, Dept. of ECE, AITS, Rajampet, A. P, India
chrajuaits@gmail.com
Shaik Karimullah
Associate Professor, Dept. of ECE, AITS, Rajampet, A. P, India
munnu483@gmail.com
Peddapullaiahgari Hariobulesu
Assitant Professor, Dept. of ECE, AITS, Rajampet, A. P, India
p. hariobulesu@gmail.com
Mohammad Sadiq S
UG Scholar, Dept. of ECE, AITS, Rajampet, A. P, India
syedsadiq5859@gmail.com

ABSTRACT

Diabetic Retinopathy (DR) is a common complication of diabetes that can lead to visual impairment and blindness if not managed effectively. In the present scenario, diabetic retinopathy remains a significant concern globally, with millions of individuals at risk of developing this condition. Early diagnosis is crucial to prevent vision loss. Traditional methods of identifying DR require manual examination by ophthalmologists, which can be time-consuming and prone to human error. This paper proposes the use of automated detection of DR model from retinal images. And also explains how the model shows difference in prediction with various sets of training and testing datasets. The model got an overall accuracy of 97% when four cases are considered where training images are 8%, 12%, 40%, and 50% and testing images are 92%, 88%, 60%, and 50% respectively. Once the model is trained and saved, it can predict whether the input fundus image is having the Diabetic Retinopathy or not. If the fundus image contains the DR it classifies the severeness of Diabetic Retinopathy.

Integrated Smart Solar Panel Management System With Tilting, Cooling, and Cleaning Using Arduino

Dr. B. Saroja
Ph. D., Professor& Head of Department ECE, Siddartha Institute of Science and Technology, India, boda. saroja@gmail.com
Vaddi Mounika
Department of ECE, Siddartha Institute of Science and Technology, India, vaddimounika758@gmail.com
Mokati Vamsikrishna
Department of ECE, Siddartha Institute of Science and Technology, India, mvamsikrishna171@gmail.com
Macherla Thilak
Department of ECE, Siddartha Institute of Science and Technology, India, macherlathilak3@gmail.com
Gandhamaneni Chushak
Department of ECE, Siddartha Institute of Science and Technology, India, saichushak1@gmail.com
Mudireddy Rajyalakshmi
Department of ECE, Siddartha Institute of Science and Technology, India, mudireddyrajyalakshmi@gmail.com

ABSTRACT

This paper presents the design and implementation of an automated cooling system for solar panels using Arduino microcontroller, temperature sensor and voltage, relay, and CPU fan. The objective of the system is to enhance the efficiency of solar panels by maintaining their temperature within an optimal range. The system utilizes a temperature sensor and voltage sensor to continuously monitor the temperature of the solar panel. The Arduino microcontroller receives temperature readings from the sensor and based on the measured temperature, the microcontroller determines whether cooling is require. When the temperature exceeds a predefined threshold, the microcontroller triggers a motor driver module to switch ON the connected to a CPU fan and pump on. The motor driver control the speed of the fan based on temperature. The fan, placed strategically near the solar panel, creates an airflow that helps dissipate heat from the panel's surface. It periodically checks the temperature and controls the fan accordingly. The ESP8266 Node MCU, acting as a Wi-Fi module, connects the system to the internet. It gathers temperature data from the Arduino and periodically uploads it to a designated web server via HTTP or other suitable protocols. This enables users to monitor temperature trends and system performance remotely, providing insights into environmental conditions and operational effectiveness. The integration of these components involves programming the Arduino and Node MCU using relevant libraries and languages such as C++ and Arduino IDE. The Arduino code manages data collection, LCD display, relay control, and fan speed regulation, ensuring seamless coordination between the components. The Node MCU code establishes a connection to the Wi-Fi network, formats and transmits data to the web server, and handles any necessary security considerations.

Keywords: Arduino, ESP8266 Node NCU, Solar panel, LDR sensor

Automatic Shuttling Metro Train Between Stations

N. Harsha Moulika
harshamoulikan@gmail.com
Mr. B. Praveen Kumar
praveenb.kumar9@gmail.com

ABSTRACT

This paper showcases advanced metro train technology used globally, featuring an autonomous system driven by a controller that eliminates the need for human operators, ensuring precision and safety. Utilizing an ATMega328P microcontroller as the CPU, the train stops automatically at stations sensed by IR sensors, facilitating passenger boarding and alighting. A door mechanism operates based on preset timings or maximum occupancy, as tracked by a passenger counting system displayed on a seven-segment display. Controlled by a motor driver IC, the train's movement is seamless and efficient. A buzzer alerts passengers during door operations and train commencement, enhancing safety measures. Future improvements could include an LCD screen displaying train status, such as expected arrival and departure times, further enhancing passenger convenience and experience. This paper encapsulates cutting-edge metro train functionalities, ensuring a smooth and automated commuting experience while paving the way for potential advancements in public transport systems.

KEY WORDS: Advanced, Autonomous, Controller, IR sensors, Passenger counting, Motor driver, Buzzer, LCD screen

An Integrated Multi-Task Feature Augmentation Approach for Scene Classification Using Remote Sensing

G. Tirumala Vasu
Department of Electronics and Communication Engineering, School of Engineering, Presidency University, Bangalore, Karnataka, India, 1sietk81nitt@gmail.com
Sandhya Tatekalva
Department of Computer Science, S. V. University, Tirupati, Andhra Pradesh, India, dr.sandhyasatish@gmail.com
Samreen Fiza
Department of Electronics and Communication Engineering, School of Engineering, Presidency University, Bangalore, Karnataka, India, samreenfiza236@gmail.com

ABSTRACT

Scene classification is critical to interpreting very high resolution (VHR) photographs. Scene categorization in remote sensing faces two challenges: the mismatching of features produced by model overfitting and the loss of semantic information. The multi-task technique solves problems by combining the share weights of many tasks. To solve these challenges, we present a feature boosting strategy based on a multi-task framework that combines the scene categorization and semantic segmentation tasks. Unlike the typical multi-task learning method, the two tasks are linked together using a weakly supervised learning method, which does not require labelled semantic segmentation examples. First, we suggested a weakly supervised segmentation approach for connecting the segmentation and classification tasks. And we get a coarse segmentation result that is highly connected with the classification using the weakly guided approach. Second, based on remote sensing's surface distribution, we suggest a sparse surface constraint to get fine segmentation results. The weakly supervised segmentation method's shared weights are constrained to produce fine features. Finally, we categorize the scenes using fine characteristics and perform tests on public remote sensing scene classification datasets. The experimental results show that the suggested linked multi-task model outperforms state-of-the-art approaches for remote sensing scene classification.

Key words: coarse segmentation, feature boosting, scene classification, semantic segmentation.

Smart Internet of Things Application for Agriculture

Dr. T. Lakshmi Narayana
Associate Professor and Head, Department of Electronics and Communication Engineering, KLM College of Engineering for Women, Kadapa, A. P., India – 516003
Email: lakshmi.svuniversity@gmail.com

Abstract

An intelligent agricultural autonomous system has been developed to regulate smart farming system application with the upward expertise called the Internet of Things (IOT). This advanced smart system carried out multitasking to enhance agri-based activities. First, monitor factors influencing the environment to increase crop productivity. Second, monitor the state of crops in the cultivation area to ensure their proper growth. Third, assess the field conditions for seasonal crops. These assessments include an investigation of a system that can track humidity, temperature, and moisture in croplands using Arduino microcontroller sensors and, in the event of a discrepancy, will send an SMS notification as well as a notification on the application developed for the same to the farmer's smartphone via Wi-Fi/4G/5G. The information that's collected by those devices is saved and analyzed in 'ThingSpeak', an IoT cloud-based platform. Because these sensors create enormous amounts of data, big data technologies are being employed for statistical analysis. The methodology created for this paper addresses these issues by providing a MapReduce architecture for managing data from sensors and other sorts of analytics for farmers to employ in smart and efficient agriculture.

Keywords— Internet of Things, IoT, Arduino Uno, ThingSpeak, data analysis, MapReduce, smart farming

An Efficient Design of Inexact Multipliers by Applying Approximations on the Partial Product Generation

P. Pradeep Kumar
Asst. Professor, Department of ECE, SRIT Proddatur
palajipradeep83@gmail.com
D. Nagaraju
Asst. Professor, Department of ECE, SRIT Proddatur
draju071@gmail.com

ABSTRACT

In modern embedded systems and data centers, energy efficiency is a mandatory design concern. Considering that a large amount of application domains exhibits an intrinsic error tolerance, e. g., digital signal processing (dsp), image processing, data analytics, and data mining, approximate computing appears as an effective solution to reduce their power dissipation. In this work approximate adders providing significant gains in terms of delay and power dissipation. Targeting the design of inexact multipliers by applying approximations on the partial product generation, we propose a novel approximate hybrid high radix encoding. The proposed technique, the most significant bits (msbs) of the multiplicand are encoded using the radix-4 encoding, whereas the k least significant bits (lsbs) are encoded using a radix-2k (with $k \geq 4$).

Approximate computing forms a design alternative that exploits the Intrinsic error Resembles of various applications and produces Energy efficient circuits with small accuracy loss. Compared with state of the art Inexact multipliers, the proposed ones constitute better approximate design alternative, Out Performing them in both

Energy Consumption and Accuracy. the proposed technique is scalable, delivering higher energy savings for the same error, as the multiplier's size increases.

Ai-Powered Healthcare: Revolutionizing Diagnosis, Treatment, and patient Care

DR. M. SubbaRao
Professor and HOD CSE, AITS, Rajampet
Y. Bhargavi
UG Scholar, AITS, Rajampet,
bhargaviyeemireddy@gmail.com
Madduru Haseena Bee
UG Scholar, AITS, Rajampet,
madduruhaseenatdp@gmail.com
N. Ankitha
UG Scholar, AITS, Rajampet
D. JayaSree
UG Scholar, AITS, Rajampet

Abstract

The abstract highlights the transformative impact of Artificial Intelligence (AI) on healthcare, with the potential to revolutionize patient diagnosis, treatment, and care. This paper delves into contemporary challenges associated with the adoption of AI-driven healthcare solutions. A primary obstacle lies in seamlessly integrating AI technologies within existing healthcare frameworks. Although AI offers the promise of improved diagnostic precision and treatment outcomes, harmonizing AI algorithms with electronic health records and clinical workflows remains a multifaceted task. Moreover, safeguarding patient data against breaches and unauthorized access is of paramount concern due to the extensive data volumes required for AI training and decision-making. The ethical dimension becomes prominent as AI algorithms take on pivotal roles in patient care decisions. Ensuring transparency, fairness, and accountability within AI systems is essential to foster trust among healthcare practitioners and patients alike. Additionally, substantial financial and logistical challenges arise from the high implementation costs of AI technologies and the necessity for comprehensive training of healthcare professionals. This paper also addresses issues concerning data quality, bias mitigation in AI algorithms, and the potential risk of excessive reliance on AI. It underscores the ongoing importance of research and development efforts to confront these challenges effectively. In conclusion, although AI presents immense potential for healthcare transformation, grappling with these challenges becomes imperative to fully realize its benefits while upholding patient safety, privacy, and ethical principles.

Implementation of Iot Based Accident Prevention and Road Safety Hardware System

V. Srivalli
4 UG Students, Department of ECE,
GMR Institute of Technology, Rajam
Y. Lokesh
UG Students, Department of ECE,
GMR Institute of Technology, Rajam
V. Gurunadh
UG Students, Department of ECE,
GMR Institute of Technology, Rajam
T. Navya
4 UG Students, Department of ECE,
GMR Institute of Technology, Rajam
Srinivas Rao
Professor, Department of ECE,
GMR Institute of Technology, Rajam
Corresponding Authors:
21341A04I9@gmrit.edu.in

ABSTRACT

In the present days, the escalating crisis of alcohol-related accidents, contributing to an annual toll of 2. 3 million premature deaths. Introducing an advanced alcohol detection system for automobile ignition, the proposed solution integrates Arduino technology. Employing a MQ3 sensor for alcohol content determination, a Micro Controller converts the output into a reading. This analysis becomes a crucial component of a comprehensive ignition locking system, preventing vehicle operation by intoxicated individuals. Emphasizing road safety, the system necessitates rolling retests to ensure ongoing sobriety during operation. By combining technological innovation with preventive measures, the proposed system strives to significantly reduce the harrowing impact of alcohol-related incidents on public health and safety, marking a vital step towards a safer and more responsible driving culture.

KEYWORDS: MQ3 alcohol sensor, ArduinoUno, relay, motor, led, buzzer.

Smart Irrigation System Using Arduino

Venu Siripurapu
UG Students, Department of ECE,
GMR Institute of Technology, Rajam
B. Vedika
UG Students, Department of ECE,
GMR Institute of Technology, Rajam
T. Uma maheswari
UG Students, Department of ECE,
GMR Institute of Technology, Rajam
V. Dilleswa rao
UG Students, Department of ECE,
GMR Institute of Technology, Rajam

ABSTRACT

smart irrigation system using Arduino is an innovative approach to optimize water usage in agriculture. This system leverages the power of Arduino microcontroller and various sensors to create an intelligent irrigation solution.

The Arduino board acts as the brain of the system, receiving inputs from sensors such as soil moisture sensors, temperature sensors, and humidity sensors. The system processes this sensor data to determine the watering needs of the plants. By considering factors like soil moisture levels, weather conditions, and plant requirements, the Arduino can intelligently control the irrigation process.

The integration of Arduino technology in this smart irrigation system brings automation, efficiency, and precision to agricultural irrigation practices. It helps farmers conserve water, save energy, enhance crop health, and ultimately increase yields.

KEYWORDS: Smart irrigation, Arduino, soil moisture sensorsirrigation valves, water pump, relay module.

Implementation of Novel System to Detect Lpg Gas Leakage and Fire Using Arduino

V. Rajeswari
UG Students, Department of ECE,
GMR Institute of Technology, Rajam, Andhra Pradesh.
V. Abhiram
UG Students, Department of ECE,
GMR Institute of Technology, Rajam, Andhra Pradesh
V. Chaitanya Sai kumar
UG Students, Department of ECE,
GMR Institute of Technology, Rajam, Andhra Pradesh
N. padmavathi
UG Students, Department of ECE,
GMR Institute of Technology, Rajam, Andhra Pradesh
Ch. Babji Prasad
Department of ECE,
GMR Institute of Technology, Rajam, Andhra Pradesh.

Abstract

Gas leakage and fire incidents pose significant risks to life and property. Rapid detection and alert systems are crucial for minimizing potential damages. In this paper, the Gas Leakage and Fire Detection System utilizing Arduino microcontroller and Liquid Crystal Display (LCD), and a buzzer is proposed. The system employs gas and fire sensors to detect hazardous situations. When either gas leakage or fire is detected, the sensors trigger the Arduino microcontroller to activate the alarm mechanism. The LCD screen displays the type of hazard detected, enabling users to take appropriate actions promptly. Additionally, the buzzer alerts nearby residents with an audible sound. This system provides real-time monitoring and alerting, enhancing safety measures in residential, commercial, and industrial settings.

Keywords : Gas leakage detection, Fire detection, Arduino UNO, Home automation, Safety alert.

Unsupervised Image Regression With Sparse Constraints in Remote Sensing Change Analysis

J. Hima bindu
Asst. Professor, Dept of ECE,
Annamacharya Institute of Technology and Sciences-Rajampet
M. Ravi Kishore
Asst. Professor, Dept of ECE,
Annamacharya Institute of Technology and Sciences-Rajampet
P. Hariobulesu
Asst. Professor, Dept of ECE,
Annamacharya Institute of Technology and Sciences-Rajampet

ABSTRACT

Change detection in remote sensing poses a significant challenge when dealing with heterogeneous multi temporal satellite images due to differing imaging mechanisms. Directly comparing such diverse images for change detection is not feasible as in the case of homogeneous images. To tackle this issue, we present an unsupervised image regression-based change detection method centered on structure consistency. In our proposed method, we begin by dynamically constructing a similarity graph to represent the structural characteristics of a pre-event image. Subsequently, this graph is used to translate the pre-event image into the domain of the post-event image, allowing us to compute the difference image. To further enhance the accuracy of change detection, we employ a super pixel-based Markovian segmentation model to classify the difference image into changed and unchanged classes. The adaptive structure consistency-based image regression model serves a dual purpose. We validate the effectiveness of our proposed method through extensive experiments on six distinct datasets, comparing its performance with several state-of-the-art methods. The results underscore the superiority of our approach in addressing the challenges posed by heterogeneous multi temporal satellite images in change detection.

Sparse Constraint Analysis With Unsupervised Image Regression

J. Hima bindu
Asst. Professor, Dept of ECE,
Annamacharya Institute of Technology and Sciences-Rajampet
M. Ravi Kishore
Asst. Professor, Dept of ECE,
Annamacharya Institute of Technology and Sciences-Rajampet
P. Hariobulesu
Asst. Professor, Dept of ECE,
Annamacharya Institute of Technology and Sciences-Rajampet

ABSTRACT

Precision of the change detection process but also contributes to the resilience of the method in the face of varying imaging mechanisms. The effectiveness of our proposed method is rigorously validated through extensive experiments conducted on six distinct datasets, each representing diverse scenarios and environments. In these experiments, we compare the performance of our approach against several state-of-the-art change detection methods. The results unequivocally underscore the superiority of our method in overcoming the challenges posed by the heterogeneous nature of multi-temporal satellite images. The method's ability to consistently outperform existing techniques showcases its potential as a reliable and robust solution for change detection in remote sensing applications. This research makes a significant contribution to the field of remote sensing by offering a novel and effective solution for change detection. Our method not only transcends the limitations of traditional approaches but also aligns with the demands of real-world applications, where accurate and timely detection of changes is crucial for tasks such as environmental monitoring and urban development analysis. The proposed approach, with its adaptive structure consistency-based regression model and super pixel-based segmentation, presents a cutting-edge methodology that can advance the capabilities of change detection in the remote sensing domain.

Mechanical Properties of Magnesium Metal Matrix Composites: Recent Developments and Applications

Nagallapati Jayakrishna
Research Scholar, School of Mechanical Engineering,
VIT, Vellore, Tamil Nadu-632014.
Dega Nagaraju
Professor, School of Mechanical Engineering,
VIT, Vellore.

Abstract:

Magnesium Metal Matrix Composites (Mg-MMCs) have gained substantial interest in both research and industrial sectors due to their exceptional mechanical properties and lightweight nature, making them ideal for applications where weight reduction and high strength are critical. This paper presents an in-depth review of the recent advancements in the development and characterization of Mg-MMCs, with a particular focus on their mechanical properties. The influence of various reinforcements, including ceramic particles, carbon nanotubes, and metallic fibers, on the tensile strength, hardness, ductility, and fatigue resistance of Mg-MMCs is thoroughly examined. Fabrication techniques such as stir casting, powder metallurgy, and friction stir processing are discussed, highlighting their impact on the microstructure and mechanical performance of the composites. Applications of Mg-MMCs in the automotive, aerospace, and biomedical industries are explored, demonstrating their potential to significantly enhance the performance and efficiency of components. The review also identifies current challenges in the production and application of Mg-MMCs and suggests future research directions to overcome these obstacles and fully exploit the benefits of these advanced materials.

Keywords: Magnesium Metal Matrix Composites, mechanical properties.

A Smart Intruder Detection System

P. Amrutha
Department of ECE,
Srinivasa Institute Of Technology And Science, Kadapa,
Mail-id- asa700008@gmail.com
N. Susmitha
Department of ECE,
Srinivasa Institute Of Technology And Science, Kadapa,
Mail-id- Nagularapususmitha@gmail.com

ABSTRACT

Presumably, the most priceless asset of a city is its inhabitants and their things. An intruder detection system is the contemporary metropolitan idea which is totally essential for inhabitants of a framework to have a quality life. This intruder detection system is used to detect an intruder and generate the alert to the authorized person. Based upon this, the incident responder can investigate the issue and take the necessary action at the instant. Here in the system the PIR sensor senses motion of human beings and it is captured by using the Pi camera. The extracted face is recognized and it is then sent to the Raspberry Pi through HDMI cable. This system is well executed in Open CV- Python. The captured image is compared with the saved image of the authorized person in the database. The system can distinguish an authorized and unauthorized person by comparing. If it is found to be an unauthorized person, then system sends the recognized image to the owner whose authorized number we feed in the system, through email using Wi-Fi technology.

Multi-Purpose Machine for Ploughing and Seeding of Agriculture Purpose Using IoT Technology

Nagallapati Jayakrishna
Assistant professor, Department of Mechanical Engineering,
Annamacharya Institute of Technology and Sciences, Rajampet,
Kadapa (Dt.), Andhra Pradesh, INDIA -516126.
jayakrishna1433@gmail.com
N bhaskar
Assistant professor, Department of Mechanical Engineering,
Annamacharya Institute of Technology and Sciences, kadapa, Kadapa (Dt.), Andhra Pradesh, INDIA -516126.
bhaskar5212@gmail.com

Abstract

In the ultramodern world, robotization is used in numerous fields similar to defense, surveillance, medical field, diligence and so on. In this design, the robot system is used to develop the process of cultivating agrarian land without the use of manpower. The end of the design is to reduce the manpower, time and increase the productivity rate. Then the designing systems like plowing the land, sowing the seed, and fertilizers scattering and navigating the vehicle stir are preferred by this independent robot using battery and DC motor. Grounded on movement of this robot in the land, the face makes position with essence distance, plough the land and drop the seeds resemblant with the help of DC Motor. The device consists of a microcontroller, which is connived with the input and affair modules, the regulator acts as an intermediate medium between both of them. The input module is nothing but a switch board to which a mobile transmitter is connected. When the stoner presses a switch, the data will be transmitted over the IOT transmitter. The data will be entered by the IOT receiver and is fed to the regulator. The Microcontroller acts consequently to program and switches the Relays to which electrical bias to be operated are connected. The Microcontroller used in the design is programmed using Embedded 'C' language.

keywords: - Arduino UNO board, solar panel, DC motor, "IOT" module.

Impact on Mechanical Properties of m30 Grade Concrete by Partial Replacement of Cement With Coal Bottom Ash and Flyash by Incorporating Alkali Activators

Dr NR Gowthami
Asst. Professor, Dept of Civil Engineering, Annamacharya Institute of Technology and Sciences-Rajampet,
nrgowthamireddy@gmail.com
Ajaya Kumar Reddy K
Asst. Professor, Dept of Mechanical Engineering, Annamacharya Institute of Technology and Sciences-Rajampet,
kakr.me@gmail.com

Abstarct

This study evaluates the mechanical properties of concrete with varying replacement ratios. The results indicate that the M3(10%flyash+15 % coal bottom ash) concrete mix exhibits superior performance, with the highest 28-day compressive strength of 36. 53 N/mm^2, and the greatest flexural strength of 3. 61 N/mm^2 among all tested mixes. Additionally, M5 (10%flyash+25 % coal bottom ash concrete achieved the highest split tensile strength of 3. 13 N/mm^2. Despite these strengths, the M3 mix did not meet the target mean strength, potentially due to insufficient reactive silica in the coal ash used, which may limit its pozzolanic activity. Furthermore, the low reactivity could be attributed to the addition of alkalis and the morphology of coal ash, which features a spherical shape and low specific surface area, affecting its performance despite the alkali environment. Overall, the study highlights that while M3(10%flyash+15 % coal bottom ash) concrete mix offers optimal replacement ratios for improved mechanical properties, its performance is constrained by material-specific factors affecting reactivity and strength development.

Predicting Remaining Useful Life of Battery Used in Electric Vehicles Using Deep Learning Algorithms

Sravanthi C L
Department of EEE, Sri Venkateswara University, Tirupathi.
sravanthi. cl@gmail.com
Chandra Sekhar J N
Department of EEE, Sri Venkateswara University, Tirupathi.
Vijaya Bhaskar K
Department of EEE, Annamacharya Institute of Technology and Sciences, Rajampet.
Sathish KC
Department of EEE, Sri Venkatesa Perumal College of Engineering and Technology, Puttur.

Abstract

The accurate prediction of the Remaining Useful Life (RUL) of batteries used in electric vehicles (EVs) is crucial for enhancing vehicle reliability, optimizing battery performance, and reducing maintenance costs. This paper investigates the application of deep learning algorithms to predict the RUL of EV batteries. Deep learning techniques, with their ability to automatically extract complex features from vast amounts of data, offer significant advantages over traditional methods. This study explores various deep learning architectures, including convolutional neural networks (CNNs), recurrent neural networks (RNNs), and long short-term memory (LSTM) networks, for RUL estimation. Using extensive datasets comprising diverse battery health indicators such as voltage, current, temperature, and state of charge, the deep learning models are trained to identify intricate patterns indicative of battery degradation. The research includes a comprehensive evaluation of these algorithms based on their prediction accuracy, computational efficiency, and robustness under different operational conditions. Real-world case studies are presented to illustrate the practical benefits and implementation of deep learning-based RUL prediction. The findings demonstrate that deep learning approaches significantly improve the accuracy and reliability of RUL predictions, enabling proactive maintenance strategies, extending battery lifespan, and enhancing the overall performance and sustainability of electric vehicles. This research underscores the potential of deep learning in advancing predictive maintenance for EV batteries and sets the stage for future innovations in this field.

Predicting Remaining Useful Life of Battery Used in Electric Vehicles Using Machine Learning Algorithms

Vijaya Bhaskar K
Department of EEE, Annamacharya Institute of Technology and Sciences, Rajampet. vijayk252@gmail.com
Senthil Kumar S
Department of EEE, Sri Venkateswara College of Engineering and Technology, Chittoor.
Murali M
Department of EEE, Vemu Institute of Technology, Pakala
Prathibha K
Department of EEE, Sri Venkateswara College of Engineering, Tirupathi

Abstract

Accurate prediction of the Remaining Useful Life (RUL) of batteries in electric vehicles (EVs) is essential for ensuring reliability, optimizing performance, and reducing maintenance costs. This paper explores the application of machine learning algorithms to predict the RUL of EV batteries. By utilizing extensive datasets comprising various battery health indicators such as voltage, current, temperature, and state of charge, machine learning models can uncover complex relationships and patterns that traditional methods might miss. This study reviews several machines learning techniques, including support vector machines, random forests, gradient boosting, and neural networks, for their effectiveness in RUL prediction. The research includes a detailed comparative analysis of these algorithms based on metrics like prediction accuracy, computational efficiency, and robustness against varying operational conditions. Additionally, the paper presents real-world case studies that demonstrate the practical benefits of machine learning-based RUL estimation for EV batteries. The results highlight that machine learning approaches significantly enhance the precision of RUL predictions, facilitating proactive maintenance strategies, extending battery lifespan, and improving the overall reliability and sustainability of electric vehicles. This research contributes to the growing body of knowledge in predictive maintenance and underscores the transformative potential of machine learning in the management of EV battery health.

Enhancing Microgrid Efficiency and Stability Through Artificial Intelligence Techniques

Vijaya Bhaskar K
Department of EEE, Annamacharya Institute of Technology and Sciences, Rajampet. vijayk252@gmail.com
Ramesh S
Department of EEE, Vel Tech Rangarajan Dr Sagunthala R&D Institute of Science and Technology, Avadi.
Venkat Rao S
Department of EEE, Vemu Institute of Technology, Pakala
Lokanadham M
Department of EEE, Sri Venkatesa Perumal College of Engineering and Technology, Puttur.

Abstract

The increasing integration of renewable energy sources and the need for resilient and efficient energy systems have led to the development of microgrids. Microgrids, being localized energy systems capable of operating independently or in conjunction with the main power grid, present unique challenges and opportunities in energy management. This paper explores the application of artificial intelligence (AI) techniques to enhance the performance, efficiency, and stability of microgrids. By leveraging advanced machine learning algorithms, predictive analytics, and optimization methods, AI can significantly improve load forecasting, energy storage management, fault detection, and real-time decision-making processes within microgrids. This study reviews current AI methodologies applied in microgrid operations, presents case studies demonstrating successful AI integration, and discusses future prospects and challenges in this rapidly evolving field. The findings indicate that AI-driven microgrids not only contribute to more reliable and sustainable energy systems but also pave the way for smarter and more adaptive energy solutions in the face of growing demand and environmental concerns.

Design and Analysis of Patient Specific Triply Peroidic Minimal Surfaces Based Models

Dr. Suresh Babu G
Assistant Professor, Department of Mechanical Engineering,
Annamacharya Institute of Technology and Sciences, Rajampet
sureshbabuhere@gmail.com
Devraj Naik
Assistant Professor, Department of Mechanical Engineering
Annamacharya Institute of Technology and Sciences, Rajampet
devraj330@gmail.com

Abstract

Accidental injuries are a typical occurrence in life, and among them are bone fractures, which can result in segmental bone abnormalities. These deformities, which are brought on by tumours, injuries, or osteoarthritis, present serious problems for the medical community and call for expensive fixes. These flaws, in contrast to some traumas, do not heal on their own and necessitate surgical intervention for bone regeneration. To replace missing bone, surgeons frequently use bone shavings from donors (allo-grafts) or from the patient's body (auto-grafts). But there are drawbacks to these approaches as well. They are linked, especially in younger patients, to aseptic loosening, periprosthetic fractures, and low survival rates. Scaffold support becomes crucial for mending segmental bone abnormalities in order to solve these problems. For the treatment of these anomalies, porous scaffolds—especially those with three-dimensional structures—offer a flexible option. When it comes to reconstructing segmental abnormalities, TPMS-based models have demonstrated potential because of their biomorphic and bio-imitating features. This project's main goal is to provide a process for building these scaffolds and evaluating how well they function under suitable loading scenarios. The scaffolds are created using additive manufacturing to fit their complex forms and geometries.

Advancements and Applications in Composite Materials: Innovations Driving Modern Engineering

P. Veera Sanjeeva Kumar
Designation:Associate professor, Department of Mechanical engineering, School of Engineering, Annamacharya University Rajampet, Andhrapradesh-516126, sanjeevpalagiri@gmail.com.
Orcid:0000-0001-6203-1767
A. Hemantha Kumar
Designation:Professor, Department of Mechanical engineering, School of Engineering, Annamacharya University Rajampet, Andhrapradesh-516126, ahk. med.aits@gmail.com.

ABSTRACT

Composite materials have revolutionized various industries by offering superior properties compared to traditional materials. This paper delves into the advancements in composite material technology, exploring the innovative methods used in their development and the diverse applications they support. The integration of fibers and matrices has led to materials that exhibit enhanced strength, durability, and lightweight characteristics, making them ideal for aerospace, automotive, civil engineering, and renewable energy sectors.

Recent progress in fabrication techniques, such as additive manufacturing and nanotechnology, has further expanded the potential of composites. This study also examines the environmental impact and sustainability of composite materials, highlighting the need for recycling and eco-friendly production methods. By understanding the current trends and future directions, this research aims to provide a comprehensive overview of composite materials and their pivotal role in shaping modern engineering solutions.

9 788197 680847

Printed at Repro India Ltd.